Suzuki 250 & 350 Twins Owners Workshop Manual

by Jeff Clew

Models covered:
T250. 247cc. 1969 to 1973
GT250. 247cc. 1973 to 1978
T305. 305cc. US only 1968
T350. 315cc. 1969 to 1973

ISBN 978 0 85696 506 7

(120-1P8)

Haynes Group Limited
Haynes North America, Inc

www.haynes.com

Acknowledgements

Grateful thanks are due to Suzuki (Great Britain) Limited for permission to reproduce their drawings. Brian Horsfall gave the necessary assistance with the dismantling and rebuilding of the 250 Suzuki Hustler used for the photographs in this manual and devised the various ingenious methods for overcoming the lack of service tools. Les Brazier arranged and took the photographs; Tim Parker edited the text and originated the layout.

Our thanks are also due to Park Motors (Yeovil) Limited, who supplied the machine photographed and to both Brian Ingle—

Finch and Louis Carr, of Huxhams (Motor Cycles) Limited, Parkstone, who provided much useful information about the overhaul and repair of the 250 cc and 350 cc Suzuki twins, based on their very extensive knowledge as Suzuki agents. We are indebted to Spalding Public Relations of London for providing the front cover transparency. Finally, we would also like to acknowledge the help of the Avon Rubber Company, who kindly supplied the illustrations that apply to tyre fitting.

About this manual

The author of this manual has the conviction that the only way in which a meaningful and easy to follow text can be written is first to do the work himself, under conditions similar to those found in the average household. As a result, the hands seen in the photographs are those of the author. Even the machine photographed was not new; an example that had covered 20,000 miles was selected so that the conditions encountered would be similar to those found by the average owner/rider. Unless specially-mentioned and therefore considered essential, Suzuki service tools have not been used. There is invariably some alternative means of loosening or slackening some vital component, when service tools are not available and risk of damage has to be avoided at all costs.

Each of the six chapters is divided into numbered sections.

Within the sections are numbered paragraphs. Cross-reference throughout this manual is quite straight forward and logical. When reference is made "See Section 6.10" - it means section 6, paragraph 10 in the same chapter. If another chapter were

meant, the text would read "See Ch apter 2, Section 6.10".

All photographs are captioned with a section/paragraph number to which they refer and are always relevant to the chapter text adjacent.

Figure numbers (usually line illustrations) appear in numerical order, within a given chapter. Fig.1.1. therefore refers to the first figure in chapter one. Left hand and right hand descriptions of the machines and their components refer to the left and right of a given machine, when the rider is seated normally.

Motorcycle manufacturers continually make changes to specifications and recommendations, and these, when notified, are incorporated into our manuals at the earliest opportunity.

Whilst every care is taken to ensure that the information in this manual is correct no liability can be accepted by the authors or publishers for loss, damage or injury caused by any errors in or omissions from the information given.

Modifications to the Suzuki 250 and 350 range

Nearly eight years have passed since the first Suzuki 250 twin reached the UK market, a period during which many design changes and detail improvements have been made. All significant changes are mentioned in the main text, under a seperate heading where appropriate. It must be appreciated that some variants of

the models included in this manual were supplied to countries other than the UK, but in the main these differences are either 'cosmetic' or relate to the lighting equipment, which has to meet the statutory requirements of the country into which the machine is imported.

Contents

1975 Suzuki GT250 model

1975 250cc Suzuki GT250A model

Introduction to the Suzuki 250cc/350cc twins

Although the Suzuki Motor Company Limited commenced manufacturing motor cycles as early as 1936, it was not until 1963 that their machines were first imported into the UK. The first of the twin cylinder models, the T10, became available during 1964 and it was immediately obvious that this particular model would be well-received by holders of provisional driving licences, who are restricted to an engine capacity limit of 250 cc. Not many 250 cc motor cycles at that period were capable of a genuine near 90 mph maximum speed and yet able to show a fuel consumption figure well in excess of 100 mpg. An electric starter and an hydraulic rear brake were additional attractions.

In 1966, the T10 was replaced by a more highly developed model known as the T20 or "Super Six". The specifications included a six-speed gearbox, hitherto virtually unknown on any standard production machine. Maximum speed was now well over 90 mph, yet it was still possible to achieve good petrol economy, albeit in the lower speed ranges. Improvements included the addition of a mechanical oil pump interconnected with the throttle, that fed oil under pressure to the crankshaft. The oil was contained in a separate oil tank, obviating the need for the earlier petroil mix. Braking benefitted from the adoption

of a large diameter twin leading shoe front brake and a manually-operated rear brake, which replaced the original hydraulic unit. Some models now have an hydraulically-operated front wheel disc brake. In order to reduce weight, the electric starter was abandoned and in its place a conventional folding kickstarter was substituted.

A larger capacity model, virtually a scaled-up T20, was introduced during 1968. This machine had an engine capacity of 305 cc and was known as the model T305. It was followed a year later by a restyled version of the T20, known as the model T250. In due course the T305 was restyled too and this became the model T350, a 315 cc version of the smaller capacity model, having a bored out 247 cc engine. Variants of the 250 cc and 350 cc models are currently available, some with added refinements such as Ram Air cooling and an hydraulic disc brake fitted to the front wheel.

Unlike other Japanese manufacturers, Suzuki have remained faithful to the two-stroke engine. Experience gained under the exacting conditions of racing has led to a World Championship, proof enough that Suzuki rank amongst the leaders in the design and manufacture of high performance two-stroke engines.

Dimensions					T250	T305	T350
Overall length (ins) ...	...	...	...	...	77.8 (1975 mm)	77.8 (1975 mm)	77.8 (1975 mm)
Overall width (ins) ...	...	...	...	...	32.2 (820 mm)	32.5 (820 mm)	32.5 (820 mm)
Overall height (ins) ...	...	...	...	...	42.5 (1080 mm)	41.9 (1065 mm)	42.5 (1080 mm)
Wheelbase (ins)	...	...	...	...	50.8 (1290 mm)	50.8 (1290 mm)	50.8 (1290 mm)
Ground clearance (ins)	...	...	...	...	6.1 (155 mm)	6.1 (155 mm)	6.3 (160 mm)
Weight (lbs) ...	...	...	...	...	308 (140 kg)	317 (144 kg)	313 (142 kg)

Ordering spare parts

When wishing to purchase spare parts for the Suzuki twins, it is best to deal direct with an accredited Suzuki agent or Suzuki (Great Britain) Limited. Either is in the best position to supply ex-stock and have more technical experience in the event of any problems that may arise. When ordering parts, always quote the frame and engine numbers IN FULL, without omitting any prefixes or suffixes. It is also advisable to include note of the colour scheme and the general styling of the machine, because there are differences between some models that use the same general coding. Avoid the use of the word "Hustler" when describing the machine, because this covers all the models in the 250 cc range. It was used only in the UK on account of its unfortunate connotations when applied in some overseas countries. To a lesser extent, this applies to the other model 'names' too.

The engine number is stamped on the left hand crankcase, to the rear of the cylinder barrels. The frame number is stamped along the right-hand side of the steering head. There is also a manufacturers' nameplate rivetted to the left-hand side of the steering head, on which the corresponding frame and engine numbers are stamped.

Always fit parts of genuine Suzuki manufacture and not pattern parts, which are often available at lower cost. Pattern parts do not necessarily make a satisfactory replacement for the originals and there are many cases where reduced life or sudden failure has occurred, to the detriment of performance.

Some of the more expendable parts such as spark plugs, bulbs, tyres, oils and greases etc., can be obtained from accessory shops and motor factors, who have convenient opening hours, charge lower prices and can often be found not far from home. It is also possible to obtain parts on a Mail Order basis from a number of specialists who advertise regularly in the motor cycle magazines.

Location of engine number

Manufacturer's name plate

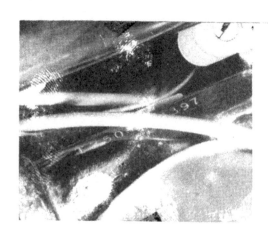

Location of frame number

Routine maintenance

Periodic routine maintenance is a continuous process that commences immediately the machine is used and continues until the machine is no longer fit for service. It must be carried out at specified mileage recordings or on a calendar basis if the machine is not used regularly, whichever the soonest. Maintenance should be regarded as an insurance policy, to help keep the machine in the peak of condition and to ensure long, trouble-free service. It has the additional benefit of giving early warning of any faults that may develop and will act as a safety check, to the obvious advantage of both rider and machine alike.

The various maintenance tasks are described, under their respective mileage and calendar headings. Accompanying diagrams are provided, where necessary. It should be remembered that the interval between the various maintenance tasks serves only as a guide. As the machine gets older, is driven hard or is used under particularly adverse conditions, it is advisable to reduce the period between each check.

Some of the tasks are described in detail, where they are not mentioned fully as a routine maintenance item in the text. If a specific item is mentioned but not described in detail, it will be covered fully in the appropriate Chapter. No special tools are required for the normal routine maintenance tasks. The tools contained in the tool kit supplied with every new machine will suffice, but if they are not available, the tools found in the average household will make an adequate substitute.

Weekly, or every 200 miles

Check the oil level through the inspection window in the side of the oil tank. If the level is close to or below the centre screw, top up with one of the prescribed oils.

Check the tyre pressures. Always check when the tyres are cold, using a pressure gauge known to be accurate.

Check the level of the electrolyte in the battery. Use only distilled water to top up, unless there has been a spillage of acid. Do not overfill.

Give the whole machine a close visual inspection, checking for loose nuts and fittings, frayed control cables etc. Make sure the lights, horn and traffic indicators function correctly, also the speedometer.

Monthly, or every 750 miles

Complete all the checks listed in the weekly/200 mile service, and then the following:

Check the operation of the oil pump. If necessary, adjust the control lever to correspond with the adjusting marks.

Clean both sparking plugs.

Adjust the play in the throttle and brake cables.

If necessary, adjust the carburettors to ensure smooth running at low rpm.

Check the contact breaker points gaps and verify whether the ignition timing is correct.

Adjust both brakes and also the amount of play in the final drive chain.

Check the tightness of the cylinder head bolts, exhaust pipe clamps and exhaust union nuts.

Check the steering head bearings for slackness.

Check both wheels for loose or broken spokes.

Three-monthly or every 2,000 miles

Complete all the checks under the weekly and monthly headings, then carry out the following additional tasks:

Remove and clean the oil tank outlet cap.

Remove, clean and lubricate the final drive chain.

Clean the air filter.

Adjust the gaps of both sparking plugs.

Change the oil in the gearbox.

Six-monthly or every 4,000 miles

Complete all the checks under the weekly, monthly and three-monthly headings, then attend to the following:

Replace both sparking plugs.

Lubricate the throttle, brake and oil pump control cables.

Decarbonise the engine and clean out the exhaust system.

Grease the twist grip throttle.

Clean the oil outlet filter and the fuel tap filter.

Yearly, or every 8,000 miles

Again complete all the checks listed under the weekly, monthly, three-monthly and six-monthly headings. The following additional tasks are now necessary:

Dismantle and clean both carburettors.

Replace both sets of contact breaker points.

Remove both wheels and check condition of front and rear brake shoes. Replace, if linings have worn thin.

RM.1 Window in oil tank shows level of contents

RM.4. Brakes are fitted with adjusters, as shown

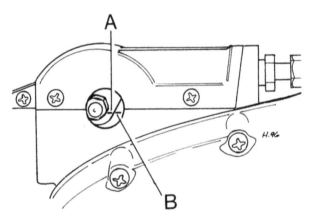

RM.2. Control lever adjusting marks for the oil pump.
(A and B must align with twist grip fully 'open').

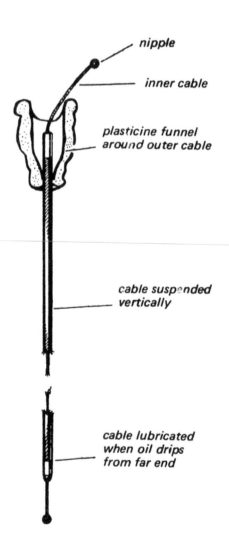

RM.5. Control cable oiling

RM.3. Cross head screw acts as level indicator when withdrawn

Quick glance
routine maintenance adjustments and capacities

Contact breaker gaps	0.012 inch – 0.016 inch	
Sparking plug gaps	0.016 inch – 0.020 inch	
Sparking plugs	NGK B–8HS or B–9H	
Fuel tank capacity	2.2 Imp gallons (T20 only) (10.0 litres)	2.6 Imp gallons all other models (11.8 litres)
Oil tank capacity	3 pints (1.8 litres)*	
Gearbox	2 pints (1.2 litres)*	

*See list of recommended lubricants for grades

Tyre pressures	Front 22 psi	Rear 26 psi	Increase rear tyre pressure only to 30 psi when a pillion passenger is carried.

Recommended lubricants

Engine:	Castrol TT two-stroke oil (1.8 litres) 3.16 Imp. pints, 3.80 U.S. pints. Note: The oil is contained in a separate oil tank, which forms part of the "Posi-Force" lubrication system. Oil SHOULD NOT be mixed with the petrol in the fuel tank.
Gearbox:	Castrol GTX (1.2 litres) 2.2 Imp. pints 2.6 U.S. pints
Grease nipples:	Castrol LM Grease
Control cables:	Castrol Everyman oil
Telescopic forks:	Castrolite (220 cc per leg) 0.39 Imp. pints 0.46 U.S. pints
Chain:	"Linklyfe" or "Chainguard"

NOTE: Capacities relate to T250/T350 models.

Safety first!

Professional motor mechanics are trained in safe working procedures. However enthusiastic you may be about getting on with the job in hand, do take the time to ensure that your safety is not put at risk. A moment's lack of attention can result in an accident, as can failure to observe certain elementary precautions.

There will always be new ways of having accidents, and the following points do not pretend to be a comprehensive list of all dangers; they are intended rather to make you aware of the risks and to encourage a safety-conscious approach to all work you carry out on your vehicle.

Essential DOs and DON'Ts

DON'T start the engine without first ascertaining that the transmission is in neutral.

DON'T suddenly remove the filler cap from a hot cooling system – cover it with a cloth and release the pressure gradually first, or you may get scalded by escaping coolant.

DON'T attempt to drain oil until you are sure it has cooled sufficiently to avoid scalding you.

DON'T grasp any part of the engine, exhaust or silencer without first ascertaining that it is sufficiently cool to avoid burning you.

DON'T allow brake fluid or antifreeze to contact the machine's paintwork or plastic components.

DON'T syphon toxic liquids such as fuel, brake fluid or antifreeze by mouth, or allow them to remain on your skin.

DON'T inhale dust – it may be injurious to health (see *Asbestos* heading).

DON'T allow any spilt oil or grease to remain on the floor – wipe it up straight away, before someone slips on it.

DON'T use ill-fitting spanners or other tools which may slip and cause injury.

DON'T attempt to lift a heavy component which may be beyond your capability – get assistance.

DON'T rush to finish a job, or take unverified short cuts.

DON'T allow children or animals in or around an unattended vehicle.

DON'T inflate a tyre to a pressure above the recommended maximum. Apart from overstressing the carcase and wheel rim, in extreme cases the tyre may blow off forcibly.

DO ensure that the machine is supported securely at all times. This is especially important when the machine is blocked up to aid wheel or fork removal.

DO take care when attempting to slacken a stubborn nut or bolt. It is generally better to pull on a spanner, rather than push, so that if slippage occurs you fall away from the machine rather than on to it.

DO wear eye protection when using power tools such as drill, sander, bench grinder etc.

DO use a barrier cream on your hands prior to undertaking dirty jobs – it will protect your skin from infection as well as making the dirt easier to remove afterwards; but make sure your hands aren't left slippery. Note that long-term contact with used engine oil can be a health hazard.

DO keep loose clothing (cuffs, tie etc) and long hair well out of the way of moving mechanical parts.

DO remove rings, wristwatch etc, before working on the vehicle – especially the electrical system.

DO keep your work area tidy – it is only too easy to fall over articles left lying around.

DO exercise caution when compressing springs for removal or installation. Ensure that the tension is applied and released in a controlled manner, using suitable tools which preclude the possibility of the spring escaping violently.

DO ensure that any lifting tackle used has a safe working load rating adequate for the job.

DO get someone to check periodically that all is well, when working alone on the vehicle.

DO carry out work in a logical sequence and check that everything is correctly assembled and tightened afterwards.

DO remember that your vehicle's safety affects that of yourself and others. If in doubt on any point, get specialist advice.

IF, in spite of following these precautions, you are unfortunate enough to injure yourself, seek medical attention as soon as possible.

Asbestos

Certain friction, insulating, sealing, and other products – such as brake linings, clutch linings, gaskets, etc – contain asbestos. *Extreme care must be taken to avoid inhalation of dust from such products since it is hazardous to health.* If in doubt, assume that they *do* contain asbestos.

Fire

Remember at all times that petrol (gasoline) is highly flammable. Never smoke, or have any kind of naked flame around, when working on the vehicle. But the risk does not end there – a spark caused by an electrical short-circuit, by two metal surfaces contacting each other, by careless use of tools, or even by static electricity built up in your body under certain conditions, can ignite petrol vapour, which in a confined space is highly explosive.

Always disconnect the battery earth (ground) terminal before working on any part of the fuel or electrical system, and never risk spilling fuel on to a hot engine or exhaust.

It is recommended that a fire extinguisher of a type suitable for fuel and electrical fires is kept handy in the garage or workplace at all times. Never try to extinguish a fuel or electrical fire with water.

Note: *Any reference to a 'torch' appearing in this manual should always be taken to mean a hand-held battery-operated electric lamp or flashlight. It does **not** mean a welding/gas torch or blowlamp.*

Fumes

Certain fumes are highly toxic and can quickly cause unconsciousness and even death if inhaled to any extent. Petrol (gasoline) vapour comes into this category, as do the vapours from certain solvents such as trichloroethylene. Any draining or pouring of such volatile fluids should be done in a well ventilated area.

When using cleaning fluids and solvents, read the instructions carefully. Never use materials from unmarked containers – they may give off poisonous vapours.

Never run the engine of a motor vehicle in an enclosed space such as a garage. Exhaust fumes contain carbon monoxide which is extremely poisonous; if you need to run the engine, always do so in the open air or at least have the rear of the vehicle outside the workplace.

The battery

Never cause a spark, or allow a naked light, near the vehicle's battery. It will normally be giving off a certain amount of hydrogen gas, which is highly explosive.

Always disconnect the battery earth (ground) terminal before working on the fuel or electrical systems.

If possible, loosen the filler plugs or cover when charging the battery from an external source. Do not charge at an excessive rate or the battery may burst.

Take care when topping up and when carrying the battery. The acid electrolyte, even when diluted, is very corrosive and should not be allowed to contact the eyes or skin.

If you ever need to prepare electrolyte yourself, always add the acid slowly to the water, and never the other way round. Protect against splashes by wearing rubber gloves and goggles.

Mains electricity and electrical equipment

When using an electric power tool, inspection light etc, always ensure that the appliance is correctly connected to its plug and that, where necessary, it is properly earthed (grounded). Do not use such appliances in damp conditions and, again, beware of creating a spark or applying excessive heat in the vicinity of fuel or fuel vapour. Also ensure that the appliances meet the relevant national safety standards.

Ignition HT voltage

A severe electric shock can result from touching certain parts of the ignition system, such as the HT leads, when the engine is running or being cranked, particularly if components are damp or the insulation is defective. Where an electronic ignition system is fitted, the HT voltage is much higher and could prove fatal.

Chapter 1 Engine, clutch and gearbox

Contents

Specifications

Model

	T250 & GT250	T305	T350
Engine	——————Twin cylinder two-stroke—————		
Cylinder heads	—————————Aluminium alloy——————————		
Cylinder barrels	—————————Aluminium alloy——————————		
Bore	54mm	59.9mm	61mm
Stroke	54mm	54mm	54mm
Cubic capacity	247cc	305cc	315cc
bhp	32 @ 8,000 rpm	37 @ 7,500 rpm	39 @ 7,500 rpm
Compression ratio (corrected)	7.5:1		6.94:1

Pistons

Type	—————Aluminium, with piston ring pegs—————
Oversizes available	+ 0.5 mm (0.020 in), + 1.0 mm (0.040 in), and + 1.5 mm (0.060 in)
	Oversize pistons for T250 and T350 models not available

Piston rings

Number	Two per piston
Gap	0.006 in - 0.014 in
Groove clearance	Not greater than 0.006 in
Capacities	
oil tank	1.8 litres
Gearbox	1.2 litres

Gear ratios

	T250 and GT250	T350
Bottom gear	20.82:1	
Second gear	13.40:1	
Third gear	10.37:1	
Fourth gear	8.04:1	
Fifth gear	6.97:1	
Top	6.34:1	

Clutch

Inserted clutch plate thickness	3.5 mm (0.138 in) standard
	3.2 mm (0.126 in) servicable limit
Clutch springs - free length	47.5 mm (1.87 in) standard
	45.5 mm (1.79 in) serviceable limit

	T250 and GT250	T305	T350
Torque wrench settings (ft lbs)			
Cylinder head nuts and bolts	14.5	14.5	14.5
Engine pinion nut	31.1	34.7	34.7
Clutch sleeve nut	23.9	23.9	23.9
Final drive sprocket nut	36.2	36.2	36 2
Transmission drain plug	23.9	23.9	23.9
Rotor bolt	14.5	14.5	14.5

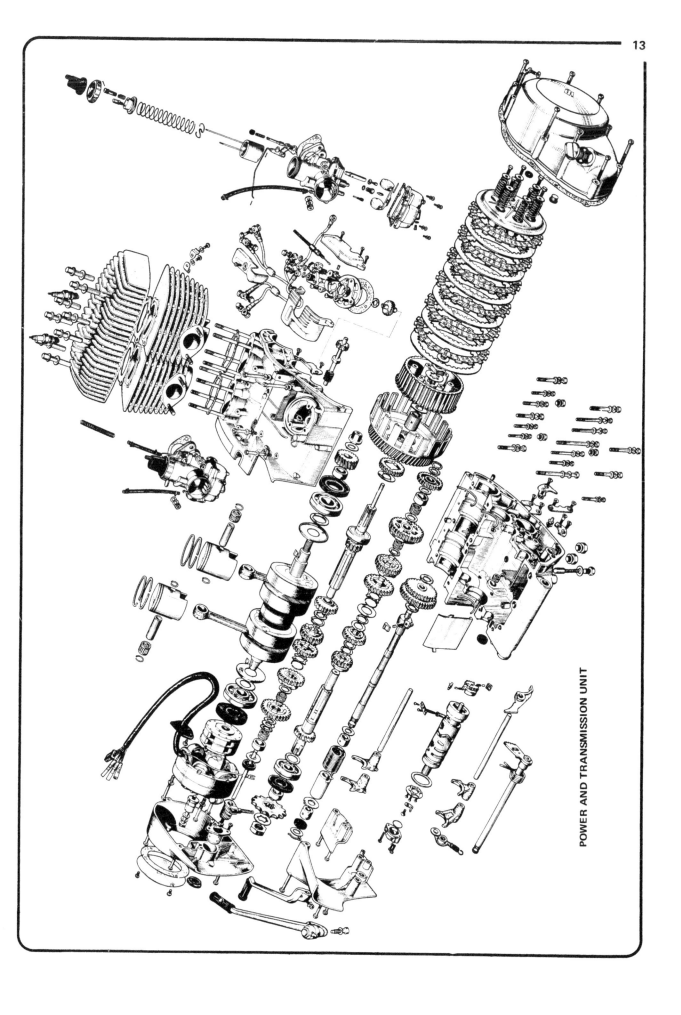

POWER AND TRANSMISSION UNIT

1 General description

The engine/gear unit fitted to the 250 cc and 350 cc Suzuki twins is of the two-stroke type employing flat top pistons and what is known as 'loop scavenging' to achieve a satisfactory induction and exhaust sequence. A rigid built-up crankshaft with thick flywheels ensures good crankcase compression; The shape and arrangement of the ports guarantees a very high standard of performance without need for mechanical aids such as rotary or reed induction valves. Another innovation is the Ram Air Cooling System. A specially-designed alloy shroud directs the air flow over the top and round the sides of the cylinder head and block, keeping engine temperature more closely under control.

The crankcase assembly is arranged to split horizontally, thereby giving maximum access to the engine and gearbox components. When the engine is dismantled, the gearbox components are fully exposed and vice-versa. It is not possible to isolate one from the other.

The gearbox has six speeds and is fitted with a conventional kickstarter. Primary drive is through a pair of helically-cut pinions, via a multi-plate clutch. A positive stop mechanism is incorporated in the gear change system, to ensure each gear is selected with a positive action.

Unlike many other two-strokes, the engine does not rely upon a petrol/oil mix for lubrication with the exception of the original T10 design. Oil is contained within a separate oil tank, from which it is fed by gravity to a mechanical oil pump interconnected with the throttle. This 'Posi-Force' system ensures that oil is delivered under pressure at all times to the crankshaft assembly and to the cylinder walls. The output of the pump is controlled by engine speed (via the throttle linkage) and thus the correct amount of oil is supplied under all operating conditions. The pump has a limited range of adjustment.

The gearbox has its own separate oil content and is fitted with a level plug to prevent accidental overfilling. It holds only a small quantity of oil, which must be changed at the recommended intervals.

2 Operations with the engine/gearbox in the frame

It is not necessary to remove the engine/gear unit from the frame unless the crankshaft assembly and/or the gearbox components require attention. Most operations can be accomplished with the engine in place, such as:

1 Removal and replacement of the cylinder heads.
2 Removal and replacement of the cylinder barrels and pistons.
3 Removal and replacement of the flywheel magneto generator.
4 Removal and replacement of the clutch.

When several operations need to be undertaken simultaneously, it would probably be an advantage to remove the complete unit from the frame, a comparatively simple operation that should take approximately twenty minutes. This will give the advantage of better access and more working space.

3 Operations with engine/gearbox removed

1 Removal and replacement of the outboard oil seals.
2 Removal and replacement of the crankshaft assembly.
3 Removal and replacement of the gear clusters, selectors and gearbox main bearings.
4 Renewal of the kickstarter return spring.

4 Method of engine/gearbox removal

As described previously, the engine and gearbox are built in

unit and it is necessary to remove the unit complete in order to gain access to either. Separation of the crankcases is accomplished after the engine unit has been removed and refitting cannot take place until the crankcases have been reassembled.

5 Removing the engine/gearbox unit

1 Place the machine on the centre stand and make sure that it is standing firmly. Remove the gearbox drain plug and drain off the oil.
2 Make sure the diaphragm-type petrol tap is in the "ON" position and pull off the two fuel pipes. Remove also the vacuum pipe that connects the left-hand carburettor flange with the petrol tap diaphragm.
3 Disconnect the electrical leads from the battery and remove the battery.
4 Disconnect the alternator wires by pulling them apart at the socket connector and the other snap connector joints. Note the wires are colour-coded to make reconnection easy.
5 Slacken off both bolts that hold the exhaust pipe union retainers in position (cross head screws) and unscrew the two unions, which have slotted ends to accept a 'C' spanner.
6 Remove the pillion footrests that act as the retainers for both silencers. The exhaust pipes complete with silencers can now be withdrawn completely.
7 Detach the clutch operating cable from the handlebar lever to gain sufficient slack for the other end to be removed. The far end can be detached after the left-hand outer engine cover has been removed. This cover contains the clutch actuating mechanism and is retained by three cross-head screws. Note that the cable end is retained by a bent-over tab.
8 Remove the final drive chain by detaching the spring link. This task is made easier if the spring link is located on the rear wheel sprocket.
9 Unscrew the tachometer drive cable connection at the top of the oil pump cover, and pull the cable end free. Remove the outer oil pump cover (2 cross head screws) and remove the throttle cable linkage.
10 Detach the flexible plastics pipe from the bottom of the oil tank by removing the union bolt. To prevent the oil content escaping from the tank, screw a 6 mm bolt into the union joint, to act as a temporary plug.
11 Pull off both sparking plug caps complete with leads and secure them away from the engine.
12 Remove both carburettor tops by unscrewing each screwed ring. The tops can be lifted away complete with their control cables, springs and throttle slide assemblies. Care is necessary at this stage to prevent damage to the slides or the needles suspended from them. Tape each top and slide assembly to a convenient frame tube, so that they are out of harm's way when the engine is lifted out.
13 Remove the air cleaner hoses from each carburettor intake. Although not strictly necessary, better access for engine removal is gained if the air filter box is removed too. This is retained at the back and the front by two small nuts and screws.
14 Take off the footrests and detach the kickstarter and the gear change lever, both of which are retained on their respective shafts by means of a pinch bolt. Remove also the cover immediately behind them, which is held in position by three cross head screws. This is the cover for the final drive gearbox sprocket.
15 Release the stop lamp spring, attached to the rear brake pedal on the right-hand side of the machine. Unscrew the rear brake adjuster and withdraw the rear brake rod from the brake operating arm. This will permit the rear brake pedal to be turned through 180°, so that it will not impede engine removal.
16 Remove the three engine bolts that hold the engine in the frame. The engine will drop in the frame unless it is supported during this operation, so a second pair of hands is advisable. Lift the engine unit upwards and to the right until it is completely clear of the frame. Note that the engine unit is heavy and that it is therefore advisable to make this operation a two-man job.

5.2. Pull fuel pipes from diaphragm tap

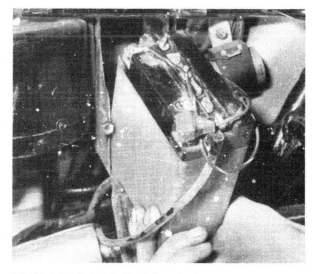

5.3. Disconnect electrical leads from battery

5.4. Some alternator wires disconnect at socket

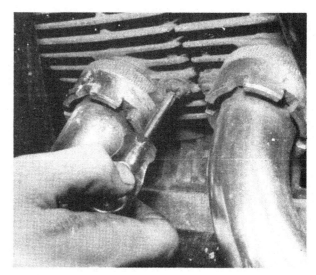

5.5. Slacken locking screws before removing exhaust union nuts

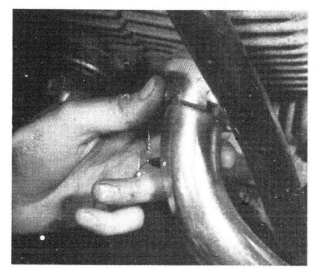

5.5a. Unions should unscrew easily after initial slackening

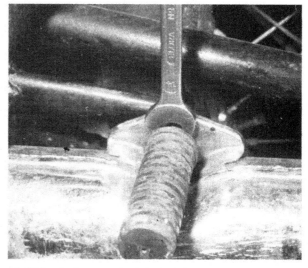

5 6. Pillion footrests retain silencers to frame

5.6a. Footrests locate with threaded lug.

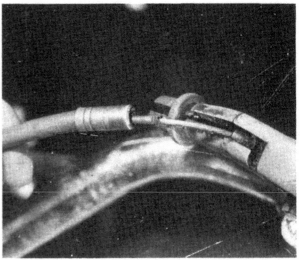

5.7. Detach clutch cable at handlebar to gain slack.....

5.7a. then remove outer engine cover and section containing cable adjuster.

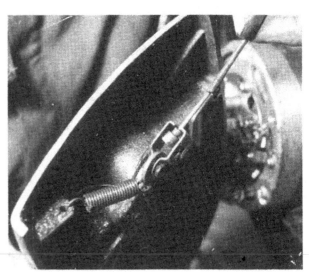

5.7b. Cable is attached to actuating mechanism within cover

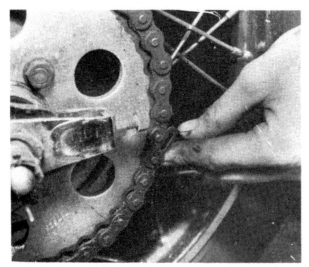

5.8. Detach spring link at rear sprocket

5.9. Unscrew tachometer drive cable

5.9a. Remove outer oil pump cover

5.9b. Detach throttle cable linkage from pump operating arm

5.9c. Nipple has slot to facilitate cable replacement

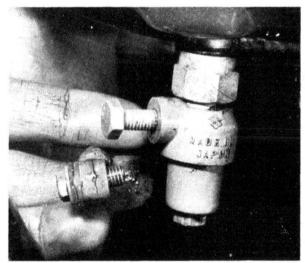

5.10. Use a 6 mm bolt as a temporary plug for oil tank union

5.12. Slides will lift out when screwed cap is removed

5.13. Clamp holds air filter hose to carburettor intake

5.13a. Engine removal is easier if air filter box is removed first

5.14. Detach footrests, held by two bolts then ......

5 14a remove kickstarter lever, held by pinch bolt

5.14b. Gear change lever is also held by pinch bolt

5.14c. Cover over final drive sprocket is retained by three screws

5.14d. will lift off when screws are removed

5.15. Stop lamp spring unhooks from brake pedal

5.16. Remove all three engine bolts

5.16a. Lift engine unit out from right hand side

6 Dismantling the engine and gearbox - general

1 Before commencing work on the engine unit, the external surfaces must be cleaned thoroughly. A motor cycle engine has very little protection from road grit and other foreign matter, which will sooner or later find its way into the dismantled engine if this simple precaution is not observed.

2 One of the proprietary engine cleaning compounds such as "Gunk" or "Jizer" can be used to good effect, especially if the compound is allowed to penetrate the film of oil and grease before it is washed away. When washing down, make sure that water cannot enter the carburettors or the electrical system, particularly if these parts are now more exposed.

3 Never use force to remove any stubborn part, unless mention is made of this requirement in the text. There is invariably good reason why a part is difficult to remove, often because the dismantling operation has been tackled in the wrong sequence.

4 Dismantling will be made easier if a simple engine stand is constructed that will correspond with the engine mounting points. This arrangement will permit the complete unit to be clamped rigidly to the work bench, leaving both hands free for the dismantling operation.

7 Dismantling the engine and gearbox - removing the carburettors

1 Each carburettor is retained to its respective cylinder barrel by two nuts and washers on the ends of holding down studs that project from the flange. Remove these nuts and washers and each carburettor can be drawn off the studs. Take care not to lose the rubber 'O' rings from the centre of each carburettor flange and the heat insulators and gaskets that will still be attached to the studs of the cylinder flange.

2 Place the carburettors aside for further attention. They are easily damaged or broken if they receive harsh treatment. If required they can be separated by withdrawing the split pin from the end of the rod that operates both chokes in unison.

8 Dismantling the engine and gearbox - removing the cylinder heads and barrels

1 Each cylinder has its own separate cylinder head. To remove each head, unscrew the cylinder head retaining bolts in a diagonal sequence, to prevent distortion.

2 When the bolts have been slackened fully and withdrawn, each cylinder head can be lifted off the long retaining studs that pass through each cylinder barrel. Note that an aluminium cylinder head gasket is used to seal the cylinder head to barrel joint. These should be discarded and not re-used, even if they appear to be in good condition.

3 Each of the separate cylinder barrels can now be lifted off by drawing them upwards along the holding down studs. Take care to support each piston as it falls clear of the cylinder barrel, otherwise the piston may be damaged or the rings broken. If only a 'top' overhaul is contemplated, it is advisable to pad the mouth of each crankcase with clean rag before the pistons are drawn clear of the cylinder barrel. This will prevent particles of broken piston ring (or displaced circlips from the next stage of the dismantling procedure) from dropping into the crankcase and causing further unnecessary dismantling.

4 Remove and discard the cylinder base gaskets, which will also have to be renewed.

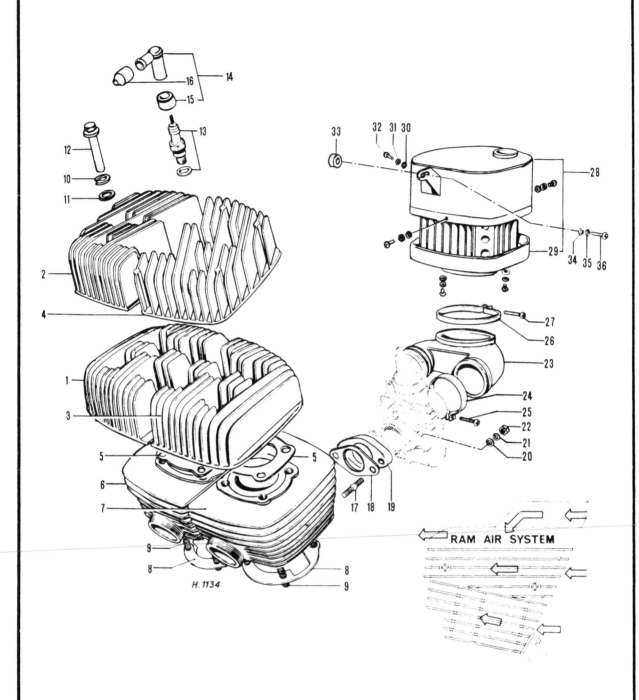

FIG.1.1. CYLINDERS AND AIR CLEANER ASSEMBLY

1	Cylinder head - right hand	9	Cylinder retaining stud - 8 off
2	Cylinder head - right hand (alternative type)	10	Spring washer - 8 off
3	Cylinder head - left hand	11	Flat washer - 8 off
4	Cylinder head left hand (alternative type)	12	Cylinder head nut - 8 off
5	Cylinder head gasket - 2 off	13	Sparking plug - 2 off
6	Cylinder - right hand	14	Sparking plug cap - 2 off
7	Cylinder - left hand	15	Sparking plug cap seal - 2 off
8	Cylinder base gasket	16	High tension lead seal-2 off

17	Carburettor mounting stud-4 off	27	Clamp screw - rear
18	Carburettor gasket - 2 off	28	Air cleaner assembly
19	Carburettor heat insulator - 2 off	29	Filter element
20	Flat washer - 4 off	30	Flat washer - 5 off
21	Spring washer - 4 off	31	Spring washer - 5 off
22	Carburettor nut - 4 off	32	Retaining screw - 2 off
23	Air cleaner hose	33	Air cleaner washer - 2 off
24	Air cleaner hose clamp - front	34	Flat washer - 2 off
25	Clamp screw - front	35	Spring washer - 2 off
26	Air cleaner hose clamp - rear	36	Retaining screw - 2 off

7.1. Release carburettors by removing retaining nuts ....

7.1a. and draw them off studs

7.1b. Remove also heat insulators and gaskets

8.1. Unscrew cylinder head bolts in a diagonal sequence

8 2. Note use of aluminium head gaskets

8.3. Ease cylinder barrels upwards along holding down studs

8.3a. Support each piston as the barrel is removed completely

5   The small end bearings take the form of caged needle rollers. Each roller assembly will lift out of the connecting rod eye.
6   Note that the piston rings are pegged so that they will remain in a set location. This is important, otherwise the rings will rotate whilst the engine is running, permitting the ends to become trapped in the ports and broken.
7   To remove the rings spread the ends sufficiently with the thumbs to allow each ring to be lifted clear of the piston. This is a very delicate operation which must be handled with great care. Piston rings are brittle and they break very easily.
8   If thes rings are stuck in their grooves or have become gummed by oily deposits, it is sometimes possible to free them by working small strips of tin along the back, to give a 'peeling' action.

8.4. Remove and discard the cylinder base gaskets

9.1. Remove and discard the circlips

9 Dismantling the engine and gearbox - removing the pistons and piston rings

1   Remove both circlips from each piston boss and discard them. Circlips should never be re-used if risk of displacement is to be obviated.
2   Using a drift of the correct diameter, tap each gudgeon pin out of the piston bosses until the piston complete with rings can be lifted off the connecting rod. Make sure the piston is properly supported during this operation, to prevent the connecting rod from bending.
3   If the gudgeon pin is a tight fit, the piston should first be warmed in order to expand the gudgeon pin bosses. A convenient way of warming the piston is to place a rag soaked in hot water on the crown.
4   When the pistons have been detached from the connecting rods, mark them on the inside of the skirt so that they will be replaced in identical positions. There is no need to mark the back and front because each piston has an arrow cast in the crown, which must always face the front of the machine.

9.2. Use drift to tap gudgeon pins out of position

9.5. Small end bearings are caged needle rollers

10 Dismantling the engine and gearbox - removing the alternator

1   Commence by removing the stator plate assembly complete with contact breakers, which is held to the left-hand side of the crankcase by three cross head screws.

2   Remove also the small plate within the outer half of the crankcase that holds the rubber grommet through which the generator leads pass. It is located in the top right-hand side of the crankcase and is retained by a single cross head screw. When the plate is removed, and the cable retaining clips slackened, the stator plate assembly complete with wiring harness can be withdrawn; there is now sufficient clearance for the terminal connector to pass through the hole in the outer half of the crankcase.

3   To remove the flywheel rotor, lock the engine by passing a stout metal rod through the eyes of the connecting rods so that it rests across the crankcase mouth. Slacken the centre bolt of the rotor (right-hand thread) and if the appropriate Suzuki extractor is not available, use a three-leg sprocket puller as shown in the accompanying photograph. Excessive force should not be necessary to break the taper joint. Tighten the sprocket puller until it is hard up against the slackened centre bolt and then give a smart hammer blow on the end of the puller. The rotor can now be withdrawn when the centre bolt is removed completely and the woodruff key placed in a safe place until it is required for reassembly. Note that the contact breaker cam will also be freed when the centre bolt is removed. Its position is predetermined by a dowel pin that engages with the rotor keyway.

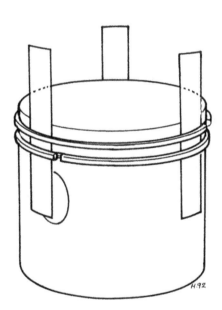

Fig. 1.2 Freeing gummed piston rings

10.1. Stator plate assembly is retained by three screws

10.2. Small plate and screw holds grommet for alternator wiring harness

10.2a. Clips retain wiring harness to top of crankcase

10.2b. Stator plate assembly is removed with wiring harness

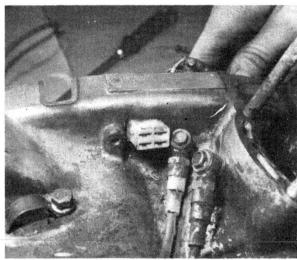

10.2c. Terminal connector will pass through hole in crankcase

10.3. Slacken rotor retaining nut after locking engine

10.3a Sprocket puller can be used to free rotor from taper

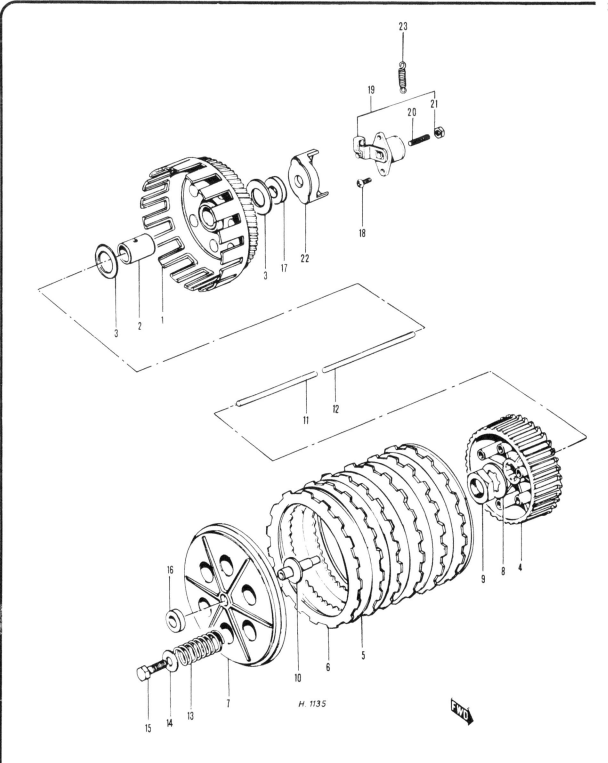

FIG. 1.3. CLUTCH

H. 1135

1	Outer drum assembly complete	
2	Spacer	
3	Thrust washer - 2 off	
4	Inner drum	
5	Plain plate - 6 off	
6	Inserted (friction) plate - 6 off	
7	Pressure plate	
8	Tab washer	
9	Clutch retaining nut	
10	Pushrod end piece	
11	Pushrod - short	
12	Pushrod - long	
13	Clutch spring - 6 off	
14	Clutch spring washer - 6 off	
15	Clutch spring bolt - 6 off	
16	Oil seal for end piece	
17	Pushrod oil seal	
18	Screw - 2 off	
19	Clutch release screw assembly	
20	Clutch release adjusting screw	
21	Locknut	
22	Clutch cover release screw	
23	Release arm spring	

11 Dismantling the engine and gearbox - removing the clutch

1 Working from the right-hand side of the engine unit, first remove the outer cover. This is held in place by ten cross head screws which are often very tight and may require the use of an impact screwdriver. Make provision for catching any surplus oil that may drain away when the cover is removed.

2 The drive gear to the oil pump (complete with spindle and worm) will lift away after the outer cover has been removed, together with the two-diameter kickstarter drive pinion. This latter pinion forms part of the kickstarter ratchet assembly and has a thrust washer at both the front and the back.

3 Lock the engine again and withdraw completely the six clutch spring bolts. The clutch pressure plate will now lift off and expose the six plain and six inserted plates that make up the clutch assembly. These plates align with the clutch inner and outer drums respectively and can be prised out of position. Lift out also the clutch pushrod end piece that fits within the hollow gearbox mainshaft.

4 Lock the engine by placing a spanner on the nut that retains the crankshaft pinion and slacken the nut that retains the inner clutch drum on the gearbox mainshaft. This nut has a right-hand thread and is retained by a tab washer. When the nut and washer have been removed from the mainshaft, the inner drum can be pulled off the shaft, followed by the outer drum and its integral driving pinion.

11.2a. Remove two-diameter kickstarter drive pinion

11.1. Removal of right hand cover will expose clutch and kickstarter mechanism

11.3. Slacken and remove all six clutch bolts

11.2. Pull out drive gear to oil pump

11.3a. Lift off pressure plate

11.3b. Clutch assembly comprises six plain and six inserted plates

11.3c. Lift out clutch pushrod end piece

11.4. Lock engine pinion in order to release clutch retaining nut

11.4a. Nut has right-hand thread and is retained by tab washer

11.4b. Inner clutch drum will pull off shaft......

11.4c. followed by outer drum and integral driving pinion

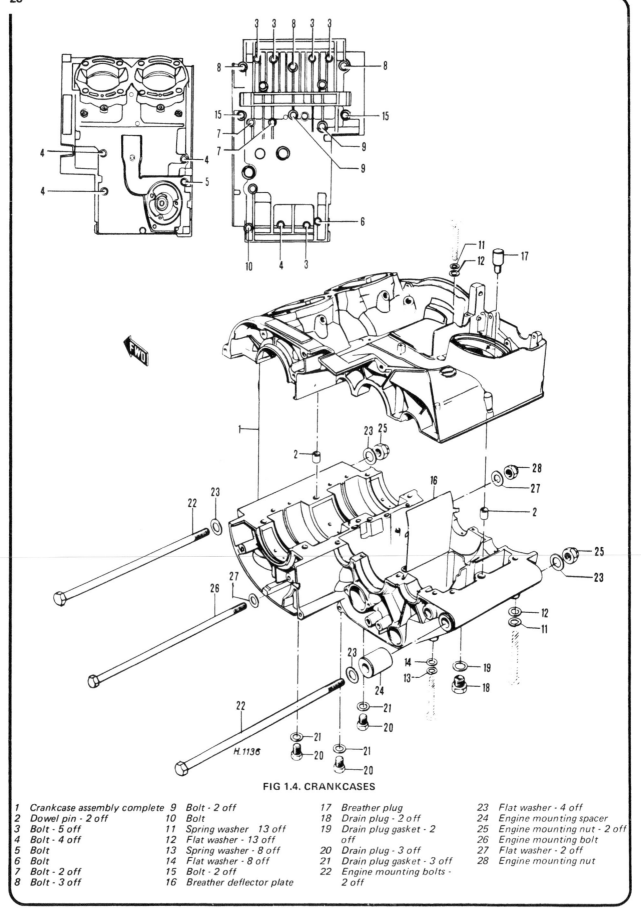

FIG 1.4. CRANKCASES

1	Crankcase assembly complete	9	Bolt - 2 off	
2	Dowel pin - 2 off	10	Bolt	
3	Bolt - 5 off	11	Spring washer 13 off	
4	Bolt - 4 off	12	Flat washer - 13 off	
5	Bolt	13	Spring washer - 8 off	
6	Bolt	14	Flat washer - 8 off	
7	Bolt - 2 off	15	Bolt - 2 off	
8	Bolt - 3 off	16	Breather deflector plate	

17	Breather plug	23	Flat washer - 4 off
18	Drain plug - 2 off	24	Engine mounting spacer
19	Drain plug gasket - 2 off	25	Engine mounting nut - 2 off
20	Drain plug - 3 off	26	Engine mounting bolt
21	Drain plug gasket - 3 off	27	Flat washer - 2 off
22	Engine mounting bolts - 2 off	28	Engine mounting nut

12 Separating the crankcases

1   Before the crankcases can be separated, a further amount of preparatory work has to be undertaken, mainly with regard to the removal of certain locking plates and part of the kickstarter and gear selector mechanism.

2   Commence by removing the kickstarter idler pinion, which is retained on the gearbox mainshaft by a circlip. Take care not to lose the two thrust washers located in front of and behind this pinion. Remove also the oil reservoir cup and the mainshaft retaining plate each of which is held in position by two counter-sunk cross head screws.

3   Pull out the gear change shaft, complete with quadrant and return spring. Note how this quadrant aligns with the gear change cam. Take off the guide plate for the gear change cam, which is situated to the immediate left of the gear change cam and the lifter for the gear change pawl, on the opposite side of the cam. Both these plates are retained by two countersunk cross head screws. The gear change cam and pawls can now be withdrawn completely from the end of the gear change drum.

4   Remove also the stopper plate for the kickstarter shaft pawl and withdraw the pawl, spring and pin.

5   Lock the engine once more and bend back the tab washer on the crankshaft pinion. Unscrew and remove the nut that retains the pinion (right-hand thread) and pull the pinion off the crank-shaft. There is a spacer behind the pinion which should be removed and a woodruff key, but no thrust washers. The crank-cases are now ready for separation.

6   Invert the engine unit and remove the various bolts that hold together the two crankcases. These total 17 in all; nine 10 mm bolts and eight 12 mm bolts. Ignore the 14 mm bolts, all of which are drain plugs!

7   Re-invert the engine and remove the four 10 mm bolts in the upper half of the engine that also retain the crankcases together. One of these is located close to the oil pump cable adjuster, which must be withdrawn before the bolt can be released.

8   A few light taps with a rawhide mallet on the projecting portions of the crankcases should ensure the crankcases separate, leaving the remainder of the engine and gearbox components in the lower half.

9   Dismantling is finished by lifting out the crankshaft assembly complete and the gear clusters. The gear selector spindles will push out of the crankcase to release the selector arms and gear selector cam stopper; the gear change drum, neutral brake, kick-starter shaft complete with return spring, and metal baffle plate will all pull clear, leaving only the oil pump attached to the uppermost crankcase.

12.2a. Remove oil reservoir cup behind pinion......

12.2b. also mainshaft retaining plate

12.2. Kickstarter idler pinion is retained by circlip

12.3. Pull out gear change shaft, complete with quadrant

12.3a. Take off guide for gear change cam.....

12.3b. also lifter for gear change pawl

12.3c. Pawls and plungers can then be lifted out

12.4. Remove stopper plate for kickstarter pawl......

12.4a. then pawl, spring and pin

12.5. Lock engine and remove nut retaining crankshaft pinion

12.5a. Retaining nut is shouldered and has a tab washer

12.9. Gear clusters will lift out after top crankcase is removed

12.9a. Also kickstarter shaft complete with return spring

12.9b. Metal plate acts as baffle for breather

12.9c. Note how gear selector pegs engage with gear change drum

12.9d. Gear selector rods pull out of crankcase, to free selector arms

12.9e. Plunger in housing actuates neutral brake and......

12.9f. ....is tensioned by compression spring

13.1. Oil pump is held to crankcase by two screws

13 1a. Spindle provides drive for both oil pump and tachometer

13 Removing the oil pump assembly

1   The oil pump is attached to a circular plate by two cross head screws. The circular plate mounts within a housing case in the uppermost crankcase and acts as the top bearing for the spindle that provides both the oil pump drive and the take-off point for the tachometer cable. This spindle has an integral pinion, which engages with the worm of the shaft that transmits the drive from the clutch via the two-diameter kickstarter pinion.

2   There is no necessity to remove the oil pump unless the complete engine/gear unit is being dismantled. The oil pump itself cannot be dismantled and should be replaced if it malfunctions in any way.

13.6. A good example of a badly worn chain

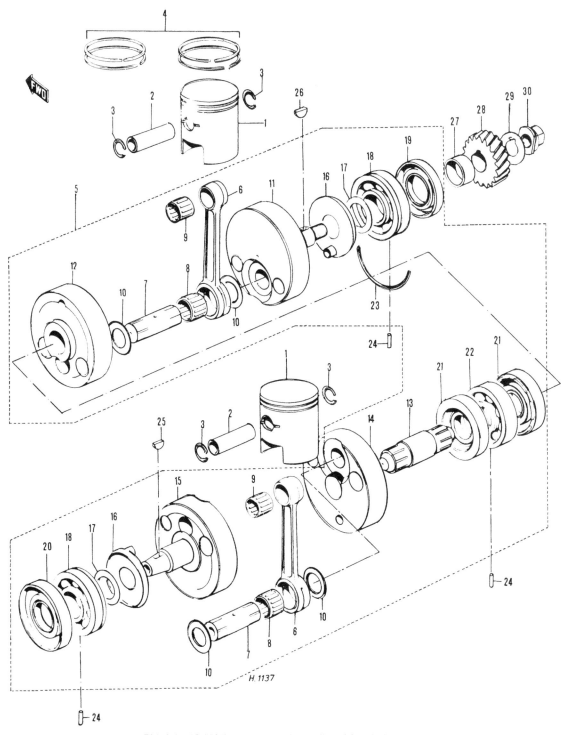

FIG.1.5. PISTONS AND CRANKSHAFT ASSEMBLY

1 Piston, right hand and left hand - 2 off	10 Connecting rod thrust washers - 4 off	17 Thrust washers - 2 off	23 'C' ring (bearing location)
2 Gudgeon pin - 2 off	11 Crankshaft - right hand	18 Crankshaft main bearings	24 Dowel pin (bearing location) - 3 off
3 Circlip - 4 off	12 Middle flywheel, right hand crankshaft	19 Oil seal - right hand crankshaft	25 Woodruff key
4 Piston ring assemblies	13 Middle crankshaft	20 Oil seal - left hand crankshaft	26 Woodruff key
5 Crankshaft assembly	14 Middle flywheel, left hand crankshaft	21 Oil seal - middle crank-shaft - 2 off	27 Crankshaft pinion spacer
6 Connecting rod - 2 off	15 Crankshaft - left hand	22 Middle crankshaft main bearing	28 Crankshaft pinion
7 Crankpin - 2 off	16 Oil guide plate - 2 off		29 Crankshaft pinion tab washer
8 Big end bearing - 2 off			30 Crankshaft pinion retaining nut
9 Small end bearing - 2 off			

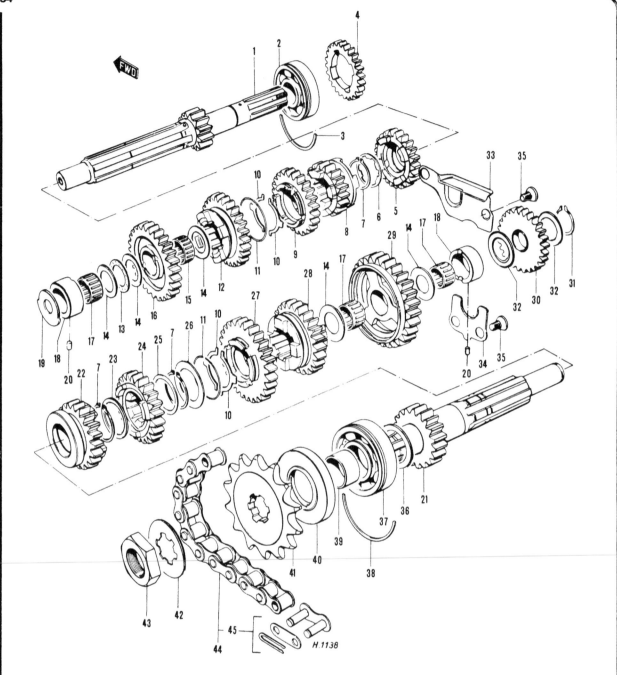

FIG 1.6. GEAR CLUSTERS

1	Layshaft	13	Top gear thrust washer - number as required	24	Mainshaft 4th gear
2	Layshaft main bearing			25	Mainshaft 4th gear thrust washer
3	'C' ring (bearing location)	14	Top gear thrust washer - number as required	26	Thrust washer
4	Kickstarter driven pinion	15	Top gear bearing	27	Mainshaft 2nd gear
5	Layshaft 3rd gear	16	Layshaft top gear	28	Mainshaft 3rd gear
6	Layshaft 3rd gear lock washer	17	Gearshaft bearing - 3 off	29	Mainshaft bottom gear
7	Circlip - 3 off	18	Gearshaft bush - 2 off	30	Kickstarter idler pinion
8	Layshaft 2nd gear	19	Layshaft retainer	31	Idler pinion circlip
9	Layshaft 4th gear	20	Dowel pin - 2 off	32	Thrust washer - 2 off
10	Knock ring - 4 off	21	Mainshaft	33	Oil reservoir cup
11	Mainshaft circlip - 2 off	22	Mainshaft 5th gear	34	Mainshaft retaining plate
12	Layshaft 5th gear	23	Thrust washer	35	Countersunk screw 4 off

36	Thrust washer
37	Mainshaft main bearing
38	'C' ring (bearing location)
39	Final drive sprocket spacer
40	Mainshaft oil seal
41	Final drive sprocket
42	Final drive sprocket tab washer
43	Final drive sprocket retaining nut
44	Final drive chain
45	Spring link

14 Removing the crankshaft and gearbox main bearings

1 Before the crankshaft outer main bearings can be removed, it is first necessary to remove the outer oil seals. These are a push fit on the crankshaft and are quite easily withdrawn. A special puller is needed to remove the outer main bearings without risk of damage to the crankshaft assembly, although it is possible to use a sprocket puller with thin jaws since there is a thrust washer between each bearing and the nearest flywheel, which provides the necessary clearance.

2 Generally speaking, it is preferable to service exchange the entire crankshaft assembly if any of the bearings are suspect. If the two outer bearings need replacing, it is highly probable that the centre bearing will be in the same condition. It is quite beyond the means of the home mechanic, or for that matter, the majority of motor cycle repair specialists, to separate and realign the crankshaft assembly without the appropriate equipment. In consequence, the Suzuki Service Exchange Scheme will provide a completely reconditioned crankshaft assembly together with new main bearings, small end bearings and oil seals for a fixed sum, in exchange for the crankshaft that requires attention. Any accredited Suzuki dealer should be able to provide this facility, often ex-stock.

3 By way of explanation, the crankshaft assembly can be regarded as two quite independent flywheel assemblies, coupled together by a 'middle' crankshaft that presses into the inner flywheels. The middle crankshaft carries the centre main bearing and also the two innermost oil seals. The two separate flywheel assemblies are built up first, aligned and checked for runout. They are then coupled together by the middle crankshaft and the entire assembly is checked for runout at both ends. It follows that a very high standard of accuracy has to be observed, in order to preserve the smooth running of the engine.

4 The gearbox assembly has only one oil seal, located immediately behind the final drive sprocket, on the mainshaft. This will pull off the shaft without difficulty. The mainshaft has a large diameter journal ball bearing at the left-hand or drive-side end and a caged needle roller bearing on the right-hand end. The reverse applies in the case of the layshaft; the journal ball bearing is on the right and the caged needle roller bearing on the left. There should be no difficulty in pulling these bearings off the shafts.

15 Dismantling the gear clusters

1 It should not be necessary to dismantle either of the gear clusters unless damage has occurred to any of the pinions or if the caged needle roller bearings require attention.

2 The accompanying illustration shows how both clusters of the six speed gearbox are assembled on their respective shafts. It is imperative that the various thrust washers are assembled in EXACTLY the correct sequence, otherwise constant gear selection problems will occur.

3 If there is any doubt about the misplacement of the various thrust washers, which look the same although they may be of varying thickness, make detailed notes or a rough sketch during the dismantling process.

16 Examination and renovation - general

1 Before examining the parts of the dismantled engine unit for wear, it is essential that they should be cleaned thoroughly. Use a paraffin/petrol mix to remove all traces of old oil and sludge that may have accumulated within the engine.

2 Examine the crankcase castings for cracks or other signs of damage. If a crack is discovered, it will require professional repair.

3 Examine carefully each part to determine the extent of wear, checking with the tolerance figures listed in the Specifications section of this Chapter. If there is any question of doubt, play

safe and renew.

4 Use a clean, lint-free rag for cleaning and drying the various components. This will obviate the risk of small particles obstructing the internal oilways, causing the lubrication system to fail.

17 Big-end and main bearings - examination and renovation

1 Failure of the big-end bearings is invariably accompanied by a knock within the crankcase that progressively becomes worse. Some vibration will also be experienced.

2 There should be no vertical play whatsoever in the big-end bearings, after the oil has been washed out. If even a small amount of vertical play is evident, the bearings are due for replacement. (A small amount of end float is both necessary and acceptable). Do not continue to run the machine with worn big-end bearings, for there is risk of breaking the connecting rods or crankshaft.

3 The built-up nature of the crankshaft assembly precludes the possibility of repair, as mentioned earlier. It will be necessary to obtain a replacement crankshaft assembly complete, under the Suzuki Service Exchange Scheme.

4 Failure of the main bearings is usually evident in the form of an audible rumble from the bottom of the engine, accompanied by vibration that is felt through the footrests.

5 The crankshaft main bearings are of the journal ball type. If wear is evident in the form of play, or if the bearings feel rough as they are rotated, replacement is necessary. Always check after the old oil has been washed out of the bearings. Whilst it is possible to remove the outer bearings at each end of the crankshaft, it is probable that the centre bearing will also require attention. Here again it will be necessary to obtain a replacement crankshaft assembly, under the Suzuki Service Exchange Scheme.

6 Failure of both the big-end bearings and the main bearings may not necessarily occur as the result of high mileage covered. If the machine is used only infrequently, it is possible that condensation within the engine may cause premature bearing failure. The condition of the flywheels is usually the best guide. When condensation troubles have occurred, the flywheels will rust and become discoloured.

18 Oil seals - examination and renovation

1 The crankshaft oil seals form one of the most critical parts in any two-stroke engine because they perform the dual function of preventing oil from leaking along the crankshaft and preventing air from leaking into the crankcase when the incoming mixture is under crankcase compression.

2 Oil seal failure is difficult to define precisely, although in most cases the machine will become difficult to start, particularly when warm. The engine will also tend to run unevenly and there will be a marked fall-off in performance, especially in the higher gears. This is caused by the intake of air into the crankcases which dilutes the mixture whilst it is under crankcase compression, giving an exceptionally weak mixture for ignition.

3 It is possible to renew the outer crankcase oil seals without difficulty, but the inner seals form part of the crankshaft assembly. In this latter case, a replacement crankshaft assembly is the only way of ensuring a satisfactory crankcase seal.

4 It is unusual for the crankcase seals to become damaged during normal service, but instances have occurred when particles of broken piston rings have fallen into the crankcases and lacerated the seals. A defect of this nature will immediately be obvious.

18.4. Damaged centre oil seals, due for replacement

19 Cylinder barrels - examination and renovation

1 The usual indication of badly worn cylinder barrels and pistons is excessive smoking from the exhausts and piston slap, a metallic rattle that occurs when there is little or no load on the engine. If the top of the bore of the cylinder barrels is examined carefully, it will be found that there is a ridge on the thrust side, the depth of which will vary according to the rate of wear that has taken place. This marks the limit of travel of the uppermost piston ring.

2 Measure the bore diameter just below the ridge, using an internal micrometer. Compare this reading with the diameter at the bottom of the cylinder bore, which has not been subjected to wear. If the difference in readings exceeds 0.05 mm (0.002 inch) the cylinder should be rebored and fitted with an oversize piston and rings.

3 If an internal micrometer is not available, the amount of cylinder bore wear can be measured by inserting the piston without rings so that it is approximately ¾ inch from the top of the bore. If it is possible to insert a 0.004 Inch feeler gauge between the piston and the cylinder wall on the thrust side of the piston, remedial action must be taken.

4 Suzuki can provide pistons in two oversizes: 0.5 mm (0.020in), and 1.0 mm (0.040in). The 1.0 mm oversize is the limit to which the cylinder can be rebored with safety.

5 Check that the surface of the cylinder bores is free from score marks or other damage that may have resulted from an earlier engine seizure or a displaced gudgeon pin. A rebore will be necessary to remove any deep indentations, irrespective of the amount of bore wear that has taken place, otherwise a compression leak will occur.

6 Make sure the external cooling fins of the cylinder barrels are not clogged with oil or road dirt, which will otherwise prevent the free flow of air and cause the engine to overheat. Remove any carbon that has accumulated in the exhaust ports, using a blunt-ended scraper so that the surface of the ports is not scratched. Finish off with metal polish so that the ports have a smooth, shiny appearance. This will aid gas flow and prevent carbon from adhering so firmly on future occasions.

7 Under no circumstances modify or re-profile the ports in the search for extra performance. The size and location of the ports is critical in terms of engine performance and the dimensions chosen have been selected to give good performance consistent with a high standard of mechanical reliability.

8 If the cylinder barrels have been rebored, it will be necessary to round off the extreme edges of the ports, to prevent rapid wear of the piston rings. Use a scraper or hand grinder and finish off with fine emery cloth.

20 Pistons and piston rings - examination and renovation

1 If a rebore is necessary, the existing pistons and piston rings can be disregarded because they will have to be replaced with their new oversize equivalents as a matter of course.

2 Remove all traces of carbon from the piston crowns, using a blunt-ended scraper to avoid scratching the surface. Finish off by polishing the crowns with metal polish, so that carbon will not adhere so readily in the future. Never use emery cloth on the soft aluminium.

3 Piston wear usually occurs at the skirt or lower end of the piston and takes the form of vertical streaks or score marks on the thrust face. There may also be some variation in the thickness of the skirt, in an extreme case.

4 The piston ring grooves may have become enlarged in use, allowing the rings to have greater side float. If the clearance exceeds 0.006 inch, the pistons are due for replacement. It is unusual for this amount of wear to occur on its own.

5 Piston ring wear is measured by removing the rings from the piston and inserting them in the cylinder, using the crown of a piston to locate them about 1½ inches from the top of the bore. Make sure they rest square in the bore. Measure the end gap with a feeler gauge; if the gap exceeds 0.040 inch the rings must be replaced (Standard gap 0.006 inch to 0.014 inch).

6 Some machines are fitted with Keystone rings, which can be identified by their shape. These rings are not interchangeable with the conventional type of rings and must be used only in conjunction with Keystone pistons. They have a 7° taper on their uppermost surface. They are handled in the same fashion as conventional rings.

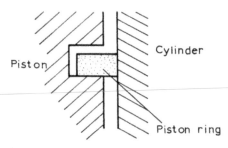

Conventional piston and Piston Ring

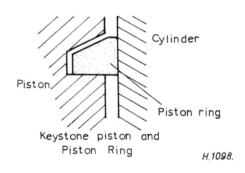

Keystone piston and Piston Ring   H.1098.

Fig.1.7. Ring profiles

21 Cylinder heads - examination and renovation

1   Remove all traces of carbon from both cylinder heads, using a blunt-ended scraper. Finish by polishing with metal polish, to give a smooth, shiny surface. This will aid gas flow and will also prevent carbon from adhering so firmly in the future.

2   Check the condition of the threads in the sparking plug holes. If the threads are worn or stretched as the result of over-tightening the plugs, they can be reclaimed by a "Helicoil" thread insert. Most dealers have the means of providing this cheap but effective repair.

3   Make sure the cylinder head fins are not clogged with oil or road dirt, otherwise the engine will overheat. If necessary, use a wire brush to clean the cooling fins. This is particularly important in the case of the Ram Air cooling heads.

4   Lay each cylinder head on a sheet of plate glass to check for distortion. Aluminium alloy cylinder heads will distort very easily, especially if the cylinder head bolts are tightened down unevenly. If the amount of distortion is only slight, it is permissible to rub the head down until it is flat once again by wrapping a sheet of very fine emery cloth around the plate glass sheet and rubbing with a rotary motion.

5   If either of the cylinder heads is distorted badly, it is advisable to fit a new replacement. Although the head joint can be restored by skimming in a lathe, this will raise the compression ratio of the engine and may adversely affect performance.

22 Gearbox components - examination and renovation

1   Give the gearbox components a close visual inspection for signs of wear or damage such as broken or chipped teeth, worn dogs, damaged or worn splines and bent selectors. Replace any parts found unserviceable because they cannot be reclaimed in a satisfactory manner.

2   Check the condition of the various pawl springs and the kickstarter return spring. If any of these springs fail after the engine has been rebuilt, a further complete stripdown of the engine unit will be necessary.

3   Check the condition of the internal serrations in the kickstarter pinion. If these serrations become worn or the edges rounded, the kickstarter will slip. Replacement of the pinion is the only means of restoring the full action.

4   Check also that the tip of the kickstarter pawl is not worn because this too will promote slip during engagement. If the kickstarter pinion is renewed, it is good policy to renew also the pawl.

5   Make sure that the pawl roller is in good shape and is free from any 'flats'. If in doubt, renew.

23 Clutch - examination and renovation

1   Cork is used as the friction material in the inserted clutch plates. When the inserts become worn, the clutch will slip, even if clutch adjustment is correct.

2   Each inserted plate should have a thickness of 3.5 mm (0.138 inch). The wear limit is 3.2 mm (0.126 inch). If the plates have reached the wear limit, they must be replaced.

3   Measure the free length of each clutch spring, because they will take a permanent set after an extended period of service. The free length of a new spring is 47.5 mm (1.87 inch). Replace ALL the clutch springs if any spring has reached the serviceable limit of 45.5 mm (1.79 inch).

4   Check also that none of the plates, either plain or inserted, is buckled. Replacement will be necessary if the plates do not lie flat, since it is difficult to straighten them with any success.

5   Check for burrs on the protruding tongues of the inserted plates and/or slots worn in the edges of the outer drum with which they engage. Wear of this nature will cause clutch drag and other troubles, since the plates will not free fully when the clutch is withdrawn. A small amount of wear can be treated by dressing with a file; more extensive wear will necessitate replacement of the parts concerned.

6   If the machine has covered a considerable mileage, it is possible that play will develop in the clutch housing, producing a mysterious rattle in the region of the clutch. This play develops in the axial direction and can be checked by pulling and pushing on the clutch. It can be reduced by grinding down one end of the spacer bush on which the clutch outer drum seats, taking care to remove only a little metal at a time.

7 Check that clutch operating mechanism in left-hand outer cover works quite freely. Mechanism operates on quick start worm principle.

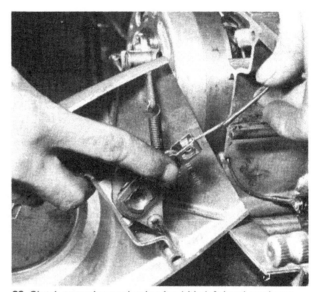

23. Clutch operating mechanism is within left-hand crankcase cover

23.7. Quick start worm operates clutch. Grease well

24 Engine and gearbox reassembly - general

1   Before reassembly is commenced, the various engine and gearbox components should be thoroughly clean and placed close to the working area.

2   Make sure all traces of old gaskets have been removed and that the mating surfaces are clean and undamaged. One of the best ways to remove old gasket cement, which is needed only on the crankcase and cover joints, is to apply a rag soaked in methylated spirit. This acts as a solvent and will ensure the cement is removed without resort to scraping and the consequent risk of damage.

3   Gather together all the necessary tools and have available an oil can filled with clean engine oil. Make sure all the new gaskets and oil seals are available; there is nothing more frustrating than having to stop in the midle of a reassembly sequence because a vital gasket or replacement has been overlooked.

4   Make sure the reassembly area is clean and well lit, with adequate working space. Refer to the torque and clearance settings wherever they are given. Many of the smaller bolts are easily sheared if they are over-tightened. Always use the correct size screwdriver bit for the cross head screws and NEVER an ordinary screwdriver or punch.

25 Engine and gearbox reassembly - rebuilding the gearbox

1   Place the lower crankcase on the workbench and commence operations by replacing the neutral brake spring and plunger.

2   Using a rawhide mallet, knock the gear selector drum and bearing into position, noting that there is a thrust washer between the crankcase and the right-hand end of the selector drum.

3   Insert the neutral brake spring and plunger, then lower gear selector arms together with their spindle, which is a push fit into the crankcase. The selector arms must engage with the tracks in the selector drum, and the shaft must pass also through the neutral brake arm, which will then be under tension.

4   Insert the upper gear selector arms together with their spindle, which is a push fit into the crankcase. This spindle carries also the stopper that engages with the cam plate on the left-hand end of the gear selector drum. Loop the end of the stopper spring over the left-hand of the two projections cast in the crankcase which retain also the breather deflector plate (to be fitted later).

5   Make sure that the upper gear selector arms are also engaged correctly with the tracks in the selector drum, then position both sets of gear clusters after ensuring the bearing retaining 'C' rings are first positioned correctly in the bearing housings.

6   Note that the gearbox bearings have a small depression in their outer faces which must locate with the dowel pins in the lower bearing housings.

7   Assembly of the gear clusters is easy if the gear selector drum is adjusted so that it is not in the neutral position. This will ensure the neutral brake is not applied. Roll the layshaft gear cluster down the already located mainshaft cluster to aid correct location. Note that a shaped retainer plate is used to anchor the left-hand end of the layshaft, which must be replaced before the two crankcases are assembled.

8   Drop the breather deflector plate into position, so that it engages with the two projections case into the crankcase. The left-hand of these two projections will have the end of the stopper spring attached.

9   Fit the kickstarter shaft assembly complete with return spring, oil seal, washer and circlip. Engage the end of the spring with the slot in the crankcase and note that the washer has a special cutaway portion to clear the end of the spring.

10 It is advisable to check that the kickstarter return spring is still in a serviceable condition since replacement at a later stage will necessitate a further complete engine strip. If in doubt, renew. A replacement spring is a low cost item. Check also that the oil seal is in good shape, to prevent oil leakage from the reassembled unit.

25 2. Do not omit thrust washer when replacing gear selector drum

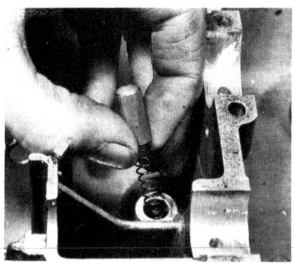

25.3. Insert neutral brake spring and plunger in housing

25.3a. Make sure selector pins engage with tracks in selector drum

25.3b. Note the upper selector shaft carries the cam plate stopper

25.7. Position mainshaft gear cluster first, then ......

25.5. Position 'C' rings in bearing housings

25.7a roll layshaft gear cluster into correct location

25.6. Note dowel pin in housing which must register with bearing

25.8. Do not forget breather deflector plate

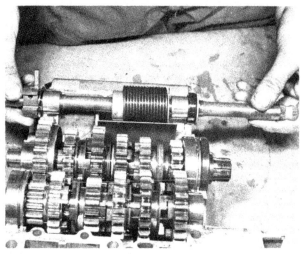

25.9. Fit kickstarter assembly with new oil seal

27.1. Crankshaft assembly should have new outer oil seals

26 Neutral brake - function and operation

1 The provision of a neutral brake is a somewhat uncommon feature in a motor cycle gearbox. An explanation of its function is therefore given.

2 The neutral brake can be regarded as a means of retarding the motion of the bottom gear pinion when neutral is selected to prevent unnecessary rotation of the various pinions that would otherwise cause gearbox noise. The braking force is applied to the layshaft when bottom gear is engaged; in other gear positions the brake arm is lifted clear and has no action.

3 The brake takes the form of a shaped arm, to which is attached a minute portion of friction lining. This lining bears on a track on the left-hand side of the layshaft bottom gear pinion when bottom gear is engaged. It is brought into engagement by a projection on the gear selector drum and is therefore fully-automatic in operation.

27 Engine and gearbox reassembly - fitting the crankshaft

1 The crankshaft assembly can be fitted to the lower crankcase immediately following reassembly of the gearbox components. The extreme right-hand main bearing is located by a 'C' ring which must be positioned first; the other two main bearings have depressions in their outer faces which must register with dowel pins in the lower bearing housings.

2 Check that the oil seals are seating correctly in their housings. It may be necessary to lightly tap the crankshaft assembly in order to ensure all components are seating correctly prior to the attachment of the upper crankcase.

28 Engine and gearbox reassembly - attaching the upper crankcase

1 Before the upper crankcase is lowered into position it is first necessary to coat the mating face of the lower crankcase with a thin layer of gasket cement. No gasket is used at this joint, in consequence a perfect seal is necessary to prevent loss of crankcase compression. Use a cement of the non-drying type, which will make any later dismantling operation much easier.

2 Lower the upper crankcase into position, making sure it registers with the two dowel pins located at the front and rear of the lower crankcase. Replace the four 10 mm bolts, one of which is close to the oil pump housing, and tighten them.

3 Invert the now complete crankcase assembly and replace the nine 10 mm bolts and the eight 12 mm bolts. A number is cast into the crankcase close to each of the bolt holes, to indicate the sequence in which they should be tightened.

27.1a. Make sure crankshaft bearings locate with dowels and 'C' rings

28.1. Check that layshaft retainer is positioned before assembling crankcases

28.1a. Lubricate the big-ends and bearings before the upper crankcase is lowered into position

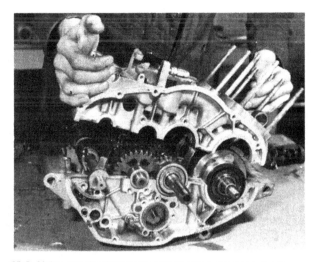

28.2. Make sure the dowel pins register when the upper crankcase is tightened down

29.1. Fit the kickstarter pawl spring and plunger......

29.1a. followed by the pawl itself

29 Engine and gearbox reassembly - replacing kickstarter and gear selector pawls

1 Tension the kickstarter shaft by temporarily fitting the kickstarter arm and depressing same. Fit the kickstarter pawl and spring, then add the stop plate, which is attached by two countersunk cross head screws. The shouldered countersunk screw fits on the right. When the stop plate is in position, the kickstarter can be released and the arm detached.

2 Assemble the two pawls and plungers of the gear change cam that locates with the right-hand end of the gear selector drum. The radiused end of the plungers must fit outwards. Insert the gear change cam whilst both pawls are depressed and retain it in position be means of the guide and stop plates, each of which is held by two countersunk cross head screws.

3 Fit also the gear change lever stop, which screws into the crankcase immediately behind the left-hand stop plate. A spring washer should be positioned between this stop and the crankcase into which it is screwed. Replace the gear change lever spindle, together with its integral quadrant and the return spring behind it. The quadrant should engage with the teeth of the gear change cam, as shown in the accompanying photograph.

4 Replace also the layshaft locking plate (2 countersunk cross head screws) and also the oil reservoir cup that is located above it and retained in similar fashion.

5 Before proceeding further, temporarily fit the gear change lever and check that the gearbox selects correctly, in all gears. It will be necessary to rotate the gear shafts by hand in order to facilitate complete engagement of the dogs.

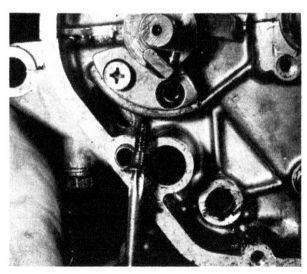

29.1b Depress the pawl and fit the stop plate, shouldered screw on right

29.2. Assemble the two pawls and plungers of the gear change cam

29.2a. Insert the gear change cam whilst the pawls are depressed then......

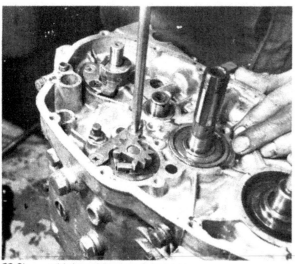

29.2b..... add the guide plate, followed by......

29.2c....... the stop plate

29.3. Replace the gear lever spindle and quadrant

29.3a. When fitted correctly, the quadrant should engage as shown

29.4. Replace the layshaft locking plate

29.4a. Also the oil reservoir cup

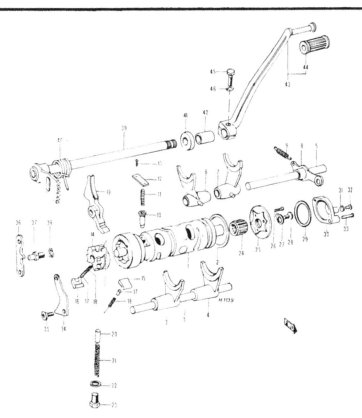

FIG.1.8. GEAR CHANGE MECHANISM AND SELECTORS

1 Gear selector drum	12 Neutral stop spring holder	25 Gear selector drum stop	39 Gear change shaft
2 Thrust washer for drum	13 Screws - 2 off	26 Screw - 3 off	40 Gear change lever return
3 Layshaft 2nd gear	14 Gear change cam	27 Neutral indicator contact	spring
selector fork	15 Gear change pawl A	28 Countersunk screw	41 Gear change shaft oil seal
4 Layshaft 5th gear	16 Gear change pawl B	29 'O' ring seal for neutral	42 Gear change shaft cushion
selector fork	17 Pawl plunger - 2 off	contact	43 Gear change lever
5 Gear selector shaft	18 Pawl spring - 2 off	30 Wipe contact for neutral	44 Rubber for gear change
- 2 off	19 Transmission brake arm	selector lamp	lever
6 Mainshaft 3rd gear	20 Transmission brake	31 Washer	45 Pinch bolt for gear change
selector fork	plunger	32 Screw	lever
7 Mainshaft 5th gear	21 Transmission brake	33 Screw - 2 off	46 Spring washer
selector fork	plunger spring	34 Gear change pawl lifter	
8 Gear selector cam	22 Transmission brake	35 Countersunk screw - 4	
stopper	plug seal	off	
9 Cam stopper spring	23 Transmission brake plug	36 Gear change cam guide	
10 Neutral stop	24 Gear selector drum	37 Gear change arm stop	
11 Neutral stop spring	bearing	38 Spring washer	

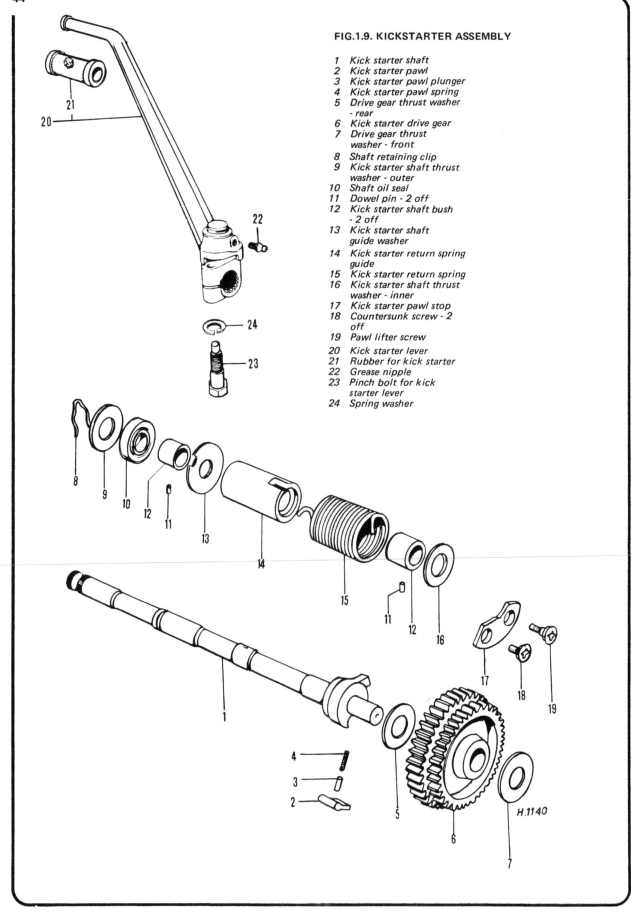

FIG.1.9. KICKSTARTER ASSEMBLY

1 Kick starter shaft
2 Kick starter pawl
3 Kick starter pawl plunger
4 Kick starter pawl spring
5 Drive gear thrust washer - rear
6 Kick starter drive gear
7 Drive gear thrust washer - front
8 Shaft retaining clip
9 Kick starter shaft thrust washer - outer
10 Shaft oil seal
11 Dowel pin - 2 off
12 Kick starter shaft bush - 2 off
13 Kick starter shaft guide washer
14 Kick starter return spring guide
15 Kick starter return spring
16 Kick starter shaft thrust washer - inner
17 Kick starter pawl stop
18 Countersunk screw - 2 off
19 Pawl lifter screw
20 Kick starter lever
21 Rubber for kick starter
22 Grease nipple
23 Pinch bolt for kick starter lever
24 Spring washer

H.1140

30 Engine and gearbox reassembly - fitting the oil pump and tachometer drive

1 Replace the circular cover plate in the rear right-hand side of the upper crankcase. Use a new gasket and make sure the worm drive pinion is inserted that provides the drive to the oil pump. There is a single thrust washer above the drive pinion. The cover is retained by two cross head screws.

2 Replace the oil pump, again using a new gasket at the joint. The projecting tongue of the oil pump drive spindle should engage with the slot in the oil pump spindle.

3 If it has been removed for access to the individual oil pipes fit also the cover that protects the oil pump junctions. This is retained by a single cross head screw.

4 Do not replace the oil pump cover at this stage, since it will be necessary to gain access to the pump union bolt when the pump is bled at a later stage.

31 Engine and gearbox reassembly - refitting the primary drive and clutch

1 Replace the idler pinion on- the gearbox mainshaft that provides part of the kickstarter drive. This pinion has thrust washers at both the front and rear and is retained on the end of the shaft by means of a circlip.

2 Replace the engine pinion after positioning the woodruff key in the end of the right-hand crankshaft. The pinion is retained by a shouldered nut and tab washer; it must be tightened fully, with the engine locked by means of a stout rod passed through both connecting rods. Bend the tab washer to secure the nut, after the latter has been tightened.

3 Replace the outer drum of the clutch on the gearbox mainshaft. It is preceded by a thrust washer and a bush which acts as a spacer. The kickstarter driven pinion must be fitted to the rear of the clutch outer drum before it is located on the splined mainshaft.

4 Fit the inner drum of the clutch, tab washer and securing nut. Lock the engine in position and tighten the centre nut fully before securing it with the tab washer. It will be necessary to 'sprag' the inner drum with the outer drum during the tightening operation, as shown in Fig. 1.10. Extreme care is necessary during this operation, to prevent damage to any of the castings.

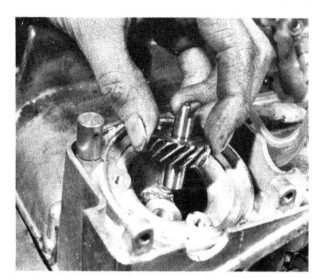

30.1. Replace the oil pump drive pinion

30.1a. Use a new gasket when replacing the oil pump mounting plate

30.2. Refit the oil pump after engaging the tongue and slot drive

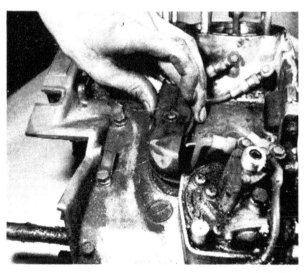

30.3. A metal cover protects the main run of the individual oil pipes

5 Fit the clutch plates, commencing with a plain metal plate that engages with the serrations of the inner drum. There are six plain and six inserted plates, which should be inserted alternately, ending with an inserted plate.

6 Fit the clutch push rod end piece, with the longer shouldered end within the hollow mainshaft. Do not omit the oil seal that locates with the other end of the end piece that projects from the shaft. It seats against the shoulder.

7 Fit the clutch pressure plate and then the six clutch springs together with their retaining bolts and washers. Tighten each bolt fully until it is firmly seated.

8 Replace in conjunction with one another the oil pump drive shaft with its integral pinion and worm, and the double diameter kickstarter drive pinion. This latter pinion has a thrust washer at both the front and rear.

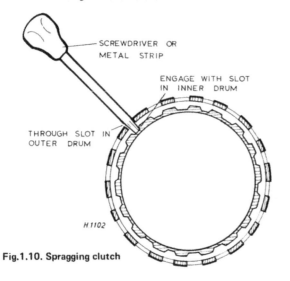

Fig.1.10. Spragging clutch

31.1. Replace the kickstarter idler pinion. Note thrust washer below

31.2. Before engine pinion is fitted, locate Woodruff key in taper

31.2a. Drive pinion on crankshaft with a drift

31.2b. Pinion is retained by tab washer and shouldered nut. Lock engine to tighten

31.3. Before clutch outer drum is fitted, thrust washer is positioned next to bearing......

31.3a followed by bush

31.3b. Lower outer clutch drum into position and

31.3c. fit second thrust washer

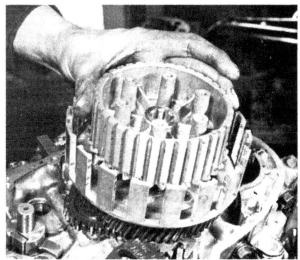
31.4. Next fit inner drum and secure by......

31.4a tab washer and. ....

31.4b. retaining nut Bend tab washer to lock nut

31.5. Fit clutch plates, commencing with plain plate

31.6. Do not forget to replace pushrod and piece before pressure plate

31.7. Lower pressure plate into position

31.7a. Add the six clutch springs and their retaining bolts and washers

31.7b. Tighten each retaining bolt fully

31.8. Replace the oil pump drive pinion and kickstarter pinion in conjunction with one another

31.8a Double diameter pinion has a thrust washer below......

32.1. Use new gasket when replacing right-hand cover

31.8b. .. ...and also above

32 Engine and gearbox reassembly - fitting the right-hand crankcase cover

1 The right-hand side of the crankcase assembly is now complete and the end cover can be added. It is located by two dowel pins at the front and rear. A gasket is used at this joint, but there is no necessity for gasket cement unless the mating surfaces are in poor condition.

2 The cover is retained by a total of 10 cross head screws, each of which should be tightened fully.

33 Engine and gearbox - fitting the neutral contact and final drive sprocket

1 Reverting to the left-hand side of the engine, fit the neutral gear electrical contact to the end of the gear selector drum. It is retained by a countersunk screw through the centre.

2 Fit the outer cover of the neutral contact, which is made of a plastic material and has an internal brass segment wiper contact. Make sure the contact surface is clean and free from oil.

3 The outer cover is fitted with the terminal uppermost. It has an internal 'O' ring seal, which must be replaced if it is in any way damaged. Tighten fully the two small cross head screws that hold the cover in position.

4 Replace the final drive sprocket on the splines of the gearbox layshaft, followed by the tab washer and the centre retaining nut. Lock the engine and tighten the centre nut fully before the tab washer is bent into the locking position.

32.1a. Two dowel pins aid correct location of cover

33.1. Neutral contact is secured in end of gear selector drum

33.3 Cover must have good 'O' ring seal

33.3a. Secure cover with two cross head screws

33.4. Final drive sprocket has splined centre to engage with shaft

34 Engine and gearbox reassembly - refitting the alternator

1   Replace the generator rotor on the left-hand end of the crankshaft, after locating the woodruff key. A taper joint fitting is used; when the rotor is located correctly, fit also the contact breaker cam and the rotor retaining bolt with its spring washer. The contact breaker cam can be fitted in only one position, as determined by the dowel pin in its base which will engage with the keyway of the rotor.

2   Tighten the rotor bolt fully, with the engine locked in position.

3   Fit the stator plate assembly by sliding it over the rotor. The stator plate is held by 3 long cross head screws that screw into the crankcase.

4   Before the stator plate assembly is slid into position, thread the wiring harness through the aperture in the crankcase housing. When the stator plate has been secured fully, the plate containing the grommet through which the wiring harness passes can be attached to the crankcase by its retaining cross head screw. This will form an effective seal to the rear of the left-hand crankcase cover when it is attached at a later stage.

5   Do not omit to attach the blue coloured wire to the terminal of the neutral contact cover.

6   Check that the ignition timing is still correct by aligning the black and red riming marks of the rotor with the line on the stator plate. A full explanation of checking and re-setting the ignition timing is given in Chapter 3, of this manual.

35 Engine and gearbox reassembly - fitting the left-hand crankcase cover

1   Before the left-hand crankcase cover is replaced, it is necessary to insert the two clutch push rods into the hollow gearbox mainshaft. They are of unequal length and the order in which they are replaced is not important.

2   It is advisable to check also the action of the clutch operating worm, the mechanism of which is attached to the inner side of the left-hand crankcase cover. The mode of operation is both simple and self-explanatory; if the operating action is harsh or uneven, the assembly should be dismantled and well greased prior to reassembly.

3   The addition of the left-hand crankcase cover at this stage is recommended only as a temporary expedient to prevent damage to the generator whilst the engine unit is being fitted in the frame of the machine. It will have to be detached again later, to insert the portion of the crankcase casting that holds the clutch cable adjuster and the cable itself.

34.1. Generator rotor is retained by bolt through the contact breaker cam

34.3. Slide stator plate assembly over rotor and......

34.3a ..... secure with three retaining screws around rim

34.4. Plate within crankcase carries wiring harness and provides seal

34.5. Blue coloured lead connects with neutral indicator contact

34.6. Timing marks show correctness of ignition timing as check

35.1. Insert new oil seal before fitting clutch pushrods

35 1a. Drop pushrods through oil seal

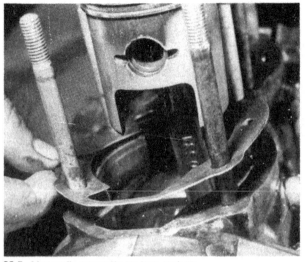

36.5. Always fit new cylinder base gaskets

36.1. Small end bearings take the form of caged needle rollers

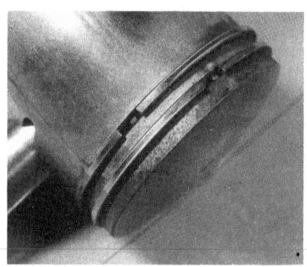

36.6. Piston rings must align with piston pegs

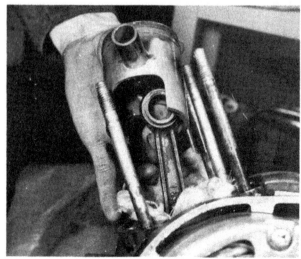

36.2. Warm piston if gudgeon pin is a tight fit

36.7. Arrow on crown must face forwards

36 Engine and gearbox - refitting the pistons and cylinder barrels

1 Before commencing assembly of the pistons and cylinder barrels, pad the mouth of each crankcase with clean rag, to prevent any misplaced component from dropping in. Replace the needle roller bearings in the connecting rod eyes.

2 Fit each piston in turn, complete with rings. If the gudgeon pins are a tight fit in the piston bosses, warm the pistons first by placing a rag soaked in hot water on the crown.

3 Use only new circlips and make sure each is positively located after the gudgeon pins have been pressed home. Never be tempted to re-use the original circlips; if they work loose as a result of having lost their tension; serious engine damage is inevitable.

4 Make sure both small ends are well lubricated by using a hand oil can filled with engine oil. Pump some into the crankshaft and big-end assemblies too, before the cylinder barrels are fitted. This will necessitate removing the rag that pads each crankcase, which should be replaced until the pistons and piston rings have engaged with each cylinder bore.

5 Before refitting each cylinder barrel, oil the bore surfaces. If the cylinders have been rebored, check to ensure the edges of the ports have been rounded off correctly. Details are given in Section 19.8 of this Chapter. Fit new base gaskets (no cement).

6 Before inserting each piston into its respective cylinder barrel, check that the piston rings are aligned correctly with regard to their end pegs. Failure to observe this precaution will lead to ring breakages, apart from extreme difficulty in fitting.

7 The pistons should be replaced in their original order, as defined by the marks scratched on the inside of the skirt during the dismantling operation. Make sure that the arrow stamped on the crown of each piston faces the forward direction ie the front of the machine.

8 There is a generous taper on the base of each cylinder barrel which should act as a good lead-in and obviate the need for a piston ring compressor. Feed each ring into the cylinder bore separately, using as little force as possible. When the pistons and rings are well up the cylinder bores, remove the rag padding from the crankcase mouths and lower the cylinder barrels until they seat firmly on their base gaskets.

9 Holding down the cylinder barrels by hand, rotate the engine a few times in both directions to make quite sure both pistons are running free in the bores. If everything appears in order, the cylinder heads can be fitted.

37 Engine and gearbox reassembly - refitting the cylinder heads

1 Place a new aluminium cylinder head gasket on each cylinder barrel and position the cylinder heads so that the sparking plug holes face the rear of the machine.

2 Fit the cylinder head bolts and tighten them down until they begin to bite. Then tighten each cylinder head separately, in a diagonal sequence, tightening each bolt a little at a time. Continue tightening in the diagonal sequence until the recommended torque wrench setting of 13 - 16.6 ft lbs is obtained.

3 Refit the sparking plugs, if only as a temporary expedient, to prevent foreign matter from dropping into the engine whilst it is being replaced. Do not over-tighten; it is sufficient to obtain a good seating between the plugs, their sealing washers and the cylinder heads, without excessive force.

38 Engine and gearbox reassembly - refitting the twin carburettors

1 Place new gaskets over each set of carburettor mounting studs followed by the heat insulators.

2 Before fitting and tightening down the carburettors, check that the 'O' ring fitted within each carburettor flange is in good condition and is seating correctly. Do not over-tighten the carburettors, otherwise the flange will bow and cause bad air leaks.

3 Join together both chokes by means of the common operating rod, which is retained by a split pin through the extreme right-hand end.

37.1. Cylinder heads have aluminium gaskets

37.2. Use two diagonally-spaced bolts for initial location

38.1. Tapered heat insulators keep carburettors cool

38.2. Do not overtighten flange nuts or flange will bow

39 Replacing the engine and gearbox unit in the frame

1 This is a task requiring the assistance of a second person because the complete engine unit is quite heavy and needs to be suported whilst the engine bolts are replaced.

2 Lift the complete unit into the frame and lower it in approximately the correct position until it rests on the lower frame tubes. It is probably easier if the engine is lifted in from the right, where there should be adequate access.

3 Support the engine unit and lift it slightly in the frame until the engine mounting lugs correspond with the engine bolt holes in the frame. Locate the rearmost engine bolt first, noting that it has a spacer close to the left-hand rear mounting plate. Fit the front engine bolt next, then the engine bolt that fits below the engine.

4 The two upper engine bolts are both of the same diameter. The lower engine bolt is smaller in diameter. It is customary to fit the engine bolts from the left-hand side of the machine, using a second person to feed them into position whilst the engine unit is lifted.

5 Each engine bolt has a washer and a nut of the self-locking type. Tighten all three bolts fully, to obviate the risk of engine vibration.

6 When the engine is secured firmly, the left-hand crankcase cover can be fitted AFTER the small portion of crankcase containing the clutch adjuster has been slid into position and the end of the cable attached to the clutch operating mechanism. It will be necessary to remove the cover before these operations can be completed if the cover has been fitted at an earlier stage as a protective measure. There is no gasket or need for gasket cement.

7 Fit also the small portion that projects from the lower part of the crankcase, which is retained by two cross head screws. The separate cover over the final drive sprocket can also be fitted, after the final drive chain has been threaded into position, and the nylon protector has been slipped over the gear change spindle. Refit kickstarter and gear change pedal.

8 Complete the fitting of the final drive chain by positioning both ends in the teeth of the rear wheel sprocket and add the spring link. Make sure the spring clip is located positively, with the closed end facing the direction of travel. Adjust the chain to the correct tension, if necessary, by following the procedure described in Chapter 5, Section 13.1. Refit the chain guard.

9 Replace both carburettor tops, with their respective slide and needle assemblies. Care is necessary to ensure the needles engage with the needle jet orifice and the slides with their grooves in line with the pegs in the carburettor body. Check that both slides operate correctly as the throttle is opened and closed; it is easy for a throttle cable to become displaced during the fitting operation.

10 Refit the air cleaner box, which is held to the frame by two cross head screws and their nuts through lugs at the front and rear. Slide the air hoses over the carburettor intakes and secure them in position with their respective clips.

11 Replace the battery and reconnect the electrical leads both to the battery and from the generator wiring harness and snap connectors. Colour coding eliminates the possibility of cross connection. The battery is retained by a strap and seats on a moulded rubber mat. Make sure the overflow pipe is connected; it is led away via a small bracket attached to the rear sub-frame so that battery acid will not damage the paintwork.

12 Replace the cover that fits over the battery housing and the electrical connections. This fits over two lugs attached to the seat tube and is held in position by a captive screwed knob. This cover contains also the tool roll, which is retained behind a flat spring clamp.

13 Replace both footrests. Note how an extension of each footrest has a separate bolt fixing to the frame, to prevent the footrests from rotating. The right-hand footrest holds the brake pedal in position.

14 Reconnect the rear brake operating cable to the brake pedal. Make sure the clevis pin is greased and a new split pin is used to retain the clevis pin in position.

15 Connect the stop lamp switch operating arm to the brake pedal by means of the spring connection. The end of the spring threads through the small hole in the top of the brake pedal arm. Check whether both the brake and the stop lamp are operating correctly and adjust either or both as necessary. The stop lamp switch has a threaded body permitting it to be either raised or lowered. The lower the switch, the later its action and vice-versa.

39.3 Rearmost engine bolt has spacer on left-hand side

39.6. Clutch adjuster slots into top of crankcase

39.6a. Left-hand crankcase cover in position

39.6b Rubber cap blanks off clutch operating mechanism

39.7. Nylon spacer protects gear change spindle from chain

39.7a. Slide end-cover into position

39.7b. Kickstarter is secured on splines by a pinch bolt

39.7c. Gear change lever fits by a similar arrangement

39.10. Air cleaner box is held to frame at front and rear

39.10a. Secure hose to bell mouth by clip connector

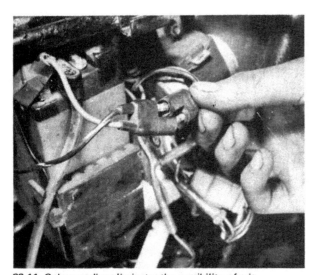

39.11. Colour coding eliminates the possibility of mis-connections

39.11a. Battery seats on rubber mat

39.11b. Tighten strap to retain battery in position

39.14. Clevis pin must have new split pin fitted

40 Reconnecting the oil pump control cable and oil pump feed

1   Returning to the right-hand side of the machine, fit the small barrel-shaped nipple over the fixing on the end of the control cable and insert this in the spring-loaded operating arm of the oil pump. The control cable is a take-off from the throttle cable junction box and passes through a cable adjuster and guide fitted close to the oil pump. A rubber sleeve fits over the cable and the adjuster, to exclude water.

2   Pump adjustment is correct when the line inscribed on the operating arm is aligned exactly with the line inscribed on the oil pump body, when the throttle is fully open. If adjustment is necessary, the cable adjuster can be used to either increase or decrease the operating arm movement. Check also that a second line inscribed on the operating arm corresponds when the throttle is closed.

3   Remove the 6 mm bolt that acts as a temporary plug to the oil feed union and reconnect the oil feed pipe that is attached to the pump. Fit new washers to the banjo union, to obviate the risk of oil leakage in service.

4   Bleed the feed pipe to the oil pump by slackening the large union nut at the forward end of the oil pump. Oil should be drained off from this point until the feed pipe is entirely free of air bubbles. Retighten the union nut, and make sure the contents of the oil tank are well above the sight glass minimum level.

5   Fit the rear cover that encases the oil pump (cross head screw) and reconnect the tachometer drive cable. Fit also the front cover (2 cross head screws) and the rubber blanking plug that seals the oil pump inspection 'window'.

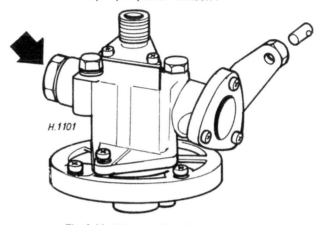

H.1101

Fig. 1.11. Union nut for oil pump bleed

40.1 Connect operating cable to oil pump arm

40.2. Marks must align when oil pump arm is in fully-open position

40.2a. Check also that marks coincide when pump arm is not under tension

40.5. Fit rear cover that encases pump and tachometer drive..

40.5a...... then fit front cover and blanking-off plug

41 Refitting the exhaust pipes and silencers

1   Before refitting the exhaust system, check that the exhaust system is clean, particularly the detachable baffles fitted to the silencers. A two-stroke engine is particularly susceptible to back pressure caused by blocked or partially blocked exhaust systems, due to the oily nature of the exhaust gases and the resultant build up of sludge.

2   Do not be tempted to replace the exhaust system without the baffles in the silencers because this will have a very adverse effect on performance. The greatly increased exhaust note may give the illusion of more speed; it is entirely false. Engine damage is liable to occur if the machine is run in this condition for an extended period.

3   When the exhaust pipes are fitted to the cylinder barrels, use new sealing washers at the port joint. Tighten the union nuts and do not omit the locking devices that engage with the nearest slot in each union nut. They are retained by a cross head screw that threads into each cylinder barrel.

4   The silencers are held to their respective frame number by the pillion footrests, which pass through lugs welded to the top of each silencer body. The right-hand silencer has an extra bracket to which is attached a rubber stop to cushion the centre stand when it is retracted.

5   Check that the screw retaining each baffle tube in the silencer ends is tight. If the screw works loose, it will eventually fall out and the baffle tubes will work free.

41.3. Always use new washers at exhaust pipe/cylinder joint

41.4. Pillion footrests secure silencers to mounting points

42 Starting and running the rebuilt engine

1 Refill the gearbox with oil of the correct viscosity until it commences to emerge from the level plug hole. Replace the filler cap.

2 Connect the petrol feed pipes to the carburettors and also the pipe from the flange of the left-hand carburettor to the diaphragm chamber of the petrol tap. This latter pipe is reinforced by an outer protective coil spring. All pipes have push-on connections.

3 Set the tap in the 'Prime' position, close the chokes and switch on the ignition. After a few depressions of the kickstarter, the engine should fire and commence to run. Raise the chokes as soon as possible, without causing the engine to stall.

4 With the engine running at between 1,500 and 2,000 rpm open the oil pump fully by pulling on the operating cable. Continue running with the oil pump in this position until all oil bubbles are expelled from the oil pipes attached to the upper crankcase. Stop the engine and return the control cable to its normal operating position, making sure the cable end is not displaced from the guide. It will be appreciated that the machine will smoke excessively whilst the engine is run with the oil pump open fully, but this is the only effective way to bleed air from the feed pipes. Excess oil may be drained from the crankcases after the engine has stopped, by removing the 14 mm drain plugs.

5 Check the exterior of the engine for signs of oil leaks or blowing gaskets. Before taking the machine on the road for the first time, check that all nuts and bolts are tight and nothing has been omitted during the reassembly sequence.

42.1 Refill gearbox with oil of correct viscosity until.....

42.1a......oil commences to emerge from level plug hole

43 Taking the rebuilt machine on the road

1 Any rebuilt engine will take time to settle down, even if parts have been replaced in their original order. For this reason it is highly advisable to treat the machine gently for the first few miles, so that oil can circulate properly and the new parts have a reasonable chance to bed down.

2 Even greater care is necessary if the engine has been rebored or if a new crankshaft has been fitted. In the case of a rebore, the engine will have to be run-in again as if the machine were new. This means more use of the gearbox and a restraining hand on the throttle until at least 500 miles have been covered. There is no point in keeping to any set speed limit; the main need is to keep only a light load on the engine and to gradually work up the performance until the 500 mile mark is reached. As a general guide, it is inadvisable to exceed 4,000 rpm during the first 500 miles and 5,000 rpm for the next 500 miles. These periods can be lessened when a new crankshaft only is fitted; experience is the best guide since it is easy to tell when the engine will run freely.

3 If at any time the oil feed shows signs of failure, stop the engine immediately and investigate the cause. If the engine is run without oil, even for a short period, irreparable engine damage is inevitable.

42.2. Pipe from tap diaphragm attaches to left-hand carburettor flange

Fault diagnosis - Engine

Symptom	Reason/s	Remedy
Engine will not start	Defective sparking plugs	Remove plugs and lay on cylinder heads. Check whether spark occurs when engine is kicked over.
	Dirty or closed contact breaker points	Check condition of points and whether gap is correct.
	Air leak at crankcases or worn crankshaft oil seals	Check whether petrol is reaching the sparking plugs (tap in prime position).
Engine runs unevenly	Ignition and/or fuel system fault	Check systems independently as though engine will not start
	Blowing cylinder head gaskets	Leak should be evident from oil leakage where gas escapes.
	Incorrect ignition timing	Check timing very accurately and reset if necessary.
Lack of power	Fault in fuel system or incorrect ignition timing	See above.
	Choked silencers	Remove and clean out baffles.
High fuel/oil consumption	Cylinder barrels in need of rebore and o/s pistons	Fit new rings and pistons after rebore.
	Oil leaks or air leaks from damaged gaskets or oil seals	Trace source of leak and replace damaged gasket and/or seal.
Excessive mechanical noise	Worn cylinder barrels (piston slap)	Rebore and fit o/s pistons
	Worn small end bearings (rattle)	Replace needle roller bearings (caged) and if necessary, gudgeon pins
	Worn big end bearings (knock)	Fit replacement crankshaft assembly.
	Worn main bearings (rumble)	Fit new journal bearings and seals. If centre bearing, new crankshaft.
Engine overheats and fades	Pre-ignition and/or weak mixture	Check carburettor settings. Check also whether plug grades correct.
	Lubrication failure	Check oil pump setting and whether oil tank is empty.

Fault diagnosis - clutch

Symptom	Reason/s	Remedy
Engine speed increases as shown by tachometer but machine does not respond	Clutch slip	Check clutch adjustment for free play at handlebar lever. Check condition of clutch plate linings, also whether clutch spring bolts are tight.
Difficulty in engaging gears. Gear changes jerky and machine creeps forward, even when clutch is withdrawn. Difficulty in selecting neutral.	Clutch drag	Check clutch adjustment for too much free play. Check for burrs on clutch plate tongues or indentations in clutch drum slots. Dress with file if damage not too great.
	Clutch assembly loose on mainshaft	Check tightness of retaining nut. If loose, fit new tab washer and retighten.
Operation action stiff	Damaged, trapped or frayed control cable	Check cable and replace if necessary. Make sure cable is lubricated and has no sharp bends.
	Bent push rods	Replace.

Fault diagnosis - gearbox

Symptom	Reason/s	Remedy
Difficulty in engaging gears	Gear selector forks bent	Replace.
	Gear clusters assembled incorrectly	Check that thrust washers are located correctly.
Machine jumps out of gear	Worn dogs on ends of gear pinions	Replace pinions involved.
	Cam plate pawls stuck	Free pawl assembly.
Gear lever does not return to normal position	Broken return spring	Replace spring.
Kickstarter does not return when engine is turned over or started	Broken or poorly tensioned return spring	Replace spring or retension.
Kickstarter slips	Kickstarter drive pinion internals, pawls or springs worn badly	Replace all worn parts.

Chapter 2 Fuel system and lubrication

Contents

Specifications

Fuel tank capacity

T250 and GT250 models	2.6 Imp. gallons (3.2 US gallons) (12 litres)
T305 models	3.1 Imp. gallons (3.7 US gallons) (14 litres)
T350 models	2.6 Imp. gallons (3.2 US gallons) (12 litres)

Oil tank capacity

All models	3.16 pints (3.80 US pints) (1.8 litres)

Carburettors

	T250 and GT250	T305	T350
Make	Mikuni	Mikuni	Mikuni
Type	VM24SH	VM32SH	VM32SH
Main jet	87.5	170	112.5
Slow running jet	30	30	30
Needle	4DH5	5DP2	5D13
Throttle valve	2.5	2.5	2.5

1 General description

1 The fuel system comprises a petrol tank from which petrol is fed by gravity to the twin carburettors, via a petrol tap of the diaphragm type. The tap is of unusual construction in the sense that it has no 'off' position. The flow of petrol is controlled by a diaphragm, which is actuated by vacuum from the left-hand carburettor flange. During each inlet cycle, the diaphragm is deflected by the vacuum effect, permitting petrol to flow to the two float chambers. When the engine stops, the diaphragm returns to its original position and the flow of petrol is shut off.

2 All models are fitted with twin carburettors of Mikuni manufacture that have integral float chambers and manually–operated chokes. The administration of the correct petrol/air mixture to the engine is controlled by a conventional throttle slide and needle arrangement. A large capacity air cleaner serves the dual purpose of supplying clean air to the carburettor intakes and effectively silencing the intake 'roar'.

3 Unlike most two-strokes of conventional design, the Suzuki twins do not depend on a petrol/oil mix for engine lubrication. They utilise the "Posi-Force" lubrication system which takes the form of a separate oil supply contained in an oil tank which is fed by gravity to an oil pump interconnected with the throttle. The oil pump provides a pressure feed to the crankshaft, big-ends and main bearings, also to the rear of each cylinder bore. The feed rate varies according to the degree of throttle opening, in a manner reminiscent of that pioneered by a British manufacturer of two-strokes during the early 1930's.

2 Petrol tank - removal and replacement

1 It is unlikely that the petrol tank will need to be removed except on very infrequent occasions, because it does not restrict access to the engine unless a top overhaul is to be carried out whilst the engine is in the frame.

2 The petrol tank is secured at the rear by a single bolt, washer and rubber buffer that threads into a strut welded across the two top frame tubes. It is necessary first to remove the dual seat before access is available.

3 When the bolt and washer are withdrawn, the petrol tank can be lifted clear from the frame. The nose of the tank is a push fit over two small rubber buffers, attached to a peg that projects from each side of the frame, immediately to the rear of the steering head. A small rubber 'mat' cushions the rear of the tank and prevents contact with the two top frame tubes.

4 The petrol tank has a locking filler cap to prevent pilferage of the tank contents when the machine is left unattended. The lock is actuated by the ignition key.

5 Unlike most two-strokes, there is no dependence on the use of a petrol/oil mix for lubrication of the engine. The engine lubrication system is quite separate and in consequence only petrol alone is carried in the petrol tank.

2.3. Rubber mat protects petrol tank from contact with frame tubes

3.1. Petrol tap is of diaphragm type and has no 'off' position

3 Petrol tap - removal, dismantling and replacement

1   As explained earlier, the petrol tap is of the diaphragm type and has no 'off' position. It is threaded into the bottom of the petrol tank on the left-hand side and has three operating positions - prime, on and reserve.

2   Before the petrol tap can be removed and dismantled, it is first necessary to drain the petrol tank. This is accomplished by turning the tap lever into the 'prime' position, when petrol will commence to flow through the tap without the engine running.

3   When the tank has been drained, unscrew the tap from the bottom of the petrol tank, using a set spanner across the flats closest to the bottom of the tank. It will be necessary to detach the three push-on pipes, one to the left-hand carburettor intake flange and one each to the carburettor float chambers, before the tap is unscrewed.

4   The filter bowl is located at the base of the tap. It is retained in position by a centre bolt, which should be removed to release the filter bowl, sealing gasket and filter gauze.

5   The diaphragm is found between the two flanges at the side of the tap assembly, held together by five small cross head screws. If these screws are removed, the diaphragm and its sealing gaskets will be exposed.

6   The petrol tap lever assembly is held to the tap body by two small screws. It is inadvisable to disturb this assembly unless leakage occurs, in which case the sealing gaskets will have to be renewed.

7   Before reassembling the tap by reversing the dismantling procedure, the condition of the diaphragm should be checked. Although the diaphragm seldom gives trouble, it has been known to rupture in isolated cases. In such cases, replacement is necessary.

8   Check also that the filter gauze is clean before the filter bowl assembly is replaced at the base of the tap. It is a wise precaution to renew the sealing washer, which may otherwise tend to leak after it has been disturbed.

9   Check also that the feed pipes and the vacuum pipe from the left-hand carburettor flange are good push-on fits, secured by their respective wire clips. The condition of the vacuum pipe is of particular importance since air leaks will seriously affect the operation of the diaphragm and the carburation of the left-hand cylinder. If in doubt, renew this pipe as a precaution; it is protected by an external coil spring throughout its length to prevent accidental damage and to provide additional reinforcement.

4 Carburettors - general description

1   The Mikuni twin carburettors fitted to the Suzuki 250/350 cc twins are of the Amal type, using a conventional throttle slide and needle arrangement that works in conjunction with a main jet to control the amount of petrol/air mixture administered to the engine. The carburettors are of identical specification, apart from the addition of a short tube to the upper portion of the left-hand carburettor flange, to which the vacuum pipe for the diaphragm petrol tap is attached.

2   Air is drawn into the carburettor bell mounths via a large capacity air cleaner that has a detachable paper element. The air cleaner acts also as an effective carburettor intake silencer and eliminates induction 'roar'. The engine must not be run without the air cleaner attached because the carburettors are jetted to compensate for the slight restriction in air flow. Removal of the air cleaner will result in a greatly weakened mixture, which will cause overheating and subsequent engine damage.

5 Carburettors - removal

1   Remove both carburettors either as a pair or separately by following the procedure described in Chapter 1, Section 7. They are linked together by the common choke-operated rod, which is freed by withdrawing the split pin through the extreme right-hand end.

2   If either or both carburettors are to be dismantled, it is advisable to remove the 'O' ring from the carburettor flange joint, to prevent it being displaced and lost.

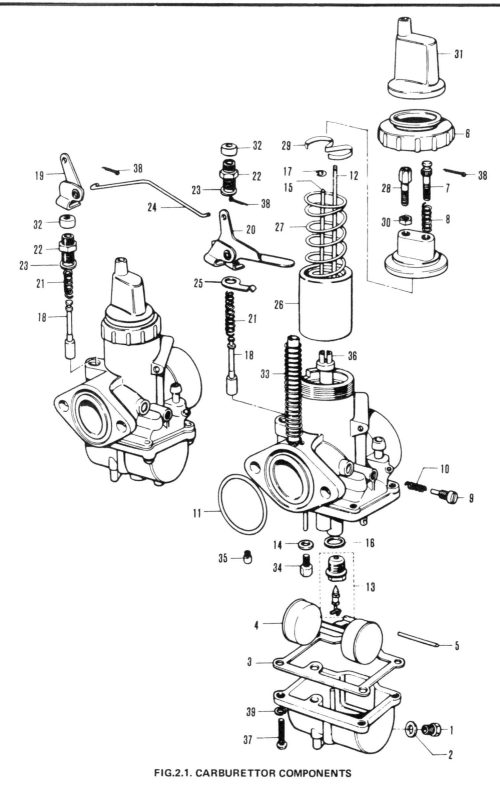

FIG.2.1. CARBURETTOR COMPONENTS

1 Drain plug
2 Drain plug gasket
3 Float chamber gasket
4 Float
5 Float pin
6 Mixing chamber cap
7 Throttle valve adjusting screw
8 Spring for throttle valve adjusting screw
9 Pilot air adjusting screw

10 Spring for pilot air adjusting screw
11 'O' ring
12 Throttle valve adjusting rod
13 Needle valve assembly
14 Needle jet stop washer
15 Jet needle
16 Needle valve seat gasket
17 Needle clip
18 Starter plunger

19 Right hand plunger lever
20 Left hand plunger lever
21 Starter plunger spring
22 Starter plunger cap
23 Starter cap washer
24 Plunger lever link
25 Starter plunger spring clip
26 Throttle valve
27 Throttle valve spring
28 Cable adjuster

29 Throttle valve spring seat
30 Cable adjuster locknut
31 Cable adjuster cap
32 Starter plunger seal washer
33 Boost tube
34 Main jet
35 Pilot jet
36 Needle jet
37 Cross head screw
38 Split pin
39 Lockwasher

6 Carburettors - dismantling and reassembly

1 Invert each carburettor and remove the float chamber by withdrawing the four retaining screws. The float chamber bowls will lift away, exposing the float assembly, hinge and float needle. There is a gasket between the float chamber bowl and the carburettor body which need not be disturbed unless it is leaking.

2 With a pair of thin nose pliers, withdraw the pin that acts as the hinge for the twin floats. This will free the floats and the float needle. Check that none of the floats has punctured and that the float needle and seating are both clean and in good condition. If the needle has a ridge, it should be renewed in conjunction with its seating.

3 Punctured floats made of brass can be repaired by soldering, after the internal petrol has been allowed to evaporate. It is questionable whether such a repair can be justified, however, other than in an emergency because the solder will add weight to the float and thus effect the petrol level. It is generally advisable to fit a new replacement float assembly in view of the comparatively low cost involved.

4 The main jet is located in the centre of the circular mixing chamber housing. It is threaded into the base of the needle jet and can be unscrewed from the bottom of the carburettor. The needle jet lifts out from the top of the carburettor, after the main jet has been unscrewed.

5 The float needle seating is also found in the underside of the carburettor, towards the bell mouth intake. It is threaded into the carburettor body and has a fibre sealing washer fitted on the underside. If the float needle and the seating are worn, they should both be replaced, never separately. Wear usually takes the form of a ridge or groove, which may cause the needle to seat imperfectly.

6 The carburettor slides, return springs and needle assemblies, together with the mixing chamber tops, are attached to the throttle cable. The throttle cable divides into two at a junction box located within the two top frame tubes. There is also a third cable, which is used to link the oil pump with the throttle.

7 After an extended period of service the throttle slides will wear and may produce a clicking sound within each carburettor body. Wear will be evident from inspection, usually at the base of the slide and in the locating groove. Worn slides should be replaced as soon as possible because they will give rise to air leaks which will upset the carburation.

8 The needles are suspended from the slides, where they are retained by a circlip. The needle is normally suspended from the centre groove, but other grooves are provided as a means of adjustment so that the mixture strength can be either increased or decreased by raising or lowering the needle. Care is necessary when replacing the carburettor tops because the needles are easily bent if they do not locate with the needle jets.

9 The manually-operated chokes are unlikely to require attention during the normal service life of the machine. When the plungers are depressed, fuel is drawn through a special starter jet in each carburettor by the partial vacuum that is created in the crankcases. Air from the float chamber passes through holes in the starter emulsion tube to aerate the fuel. The fuel then mixes with air drawn in via the starter air inlet to the plunger chamber. The resultant mixture, enriched for a cold start, is drawn into the engine through the starter outlet, behind the throttle valve. The accompanying diagram shows the location of the various choke system components and their relationship with the other carburettor components.

10 Before the carburettors are reassembled, using the reversed dismantling procedure, each should be cleaned out thoroughly, preferably by the use of compressed air. Avoid using a rag because there is always risk of fine particles of lint obstructing the internal air passages or the jet orifices.

11 Never use a piece of wire or any sharp metal object to clear a blocked jet. It is only too easy to enlarge the jet under these circumstances and increase the rate of petrol consumption. Always use compressed air to clear a blockage; a tyre pump

makes an admirable substitute when a compressed air line is not available.

12 Do not use excessive force when reassembling the carburettors because it is quite easy to shear the small jets or some of the smaller screws. Before attaching the air cleaner hoses, check that both throttle slides rise when the throttle is opened.

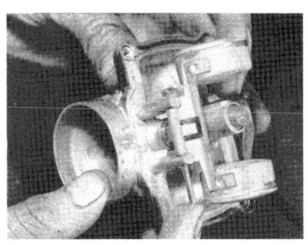

6.1. Removal of float chamber exposes floats

6.1a. Note gasket between carburettor body and float chamber

6.4. Needle jet lifts from top of carburettor after main jet is removed

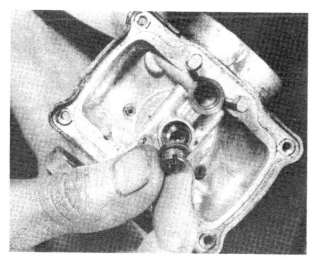

6.4a. Main jet screws into base of needle jet

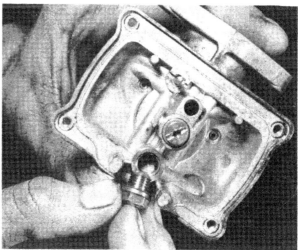

6.5a. Needle seating threads into base of carburettor body

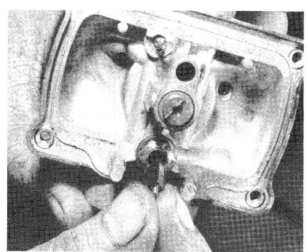

6.5. Float needle locates with seating, as shown

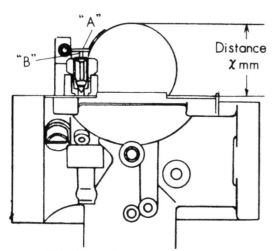

FIG.2.2. CHECKING FLOAT LEVEL

A – Float tongue
B – Float valve
X – 25.7 mm (1 in.) T250
    27.5 mm (1.1 in.) T350

7 Carburettors - adjustment

1 When adjustments are required it is best to regard each carburettor as a separate entity. Remove the sparking plug from the cylinder that is not receiving attention so that only the one cylinder that is fire during the adjustments. Then change over and follow a similar routine. These adjustments should be made with the engine warm.

2 Commence operations by checking the float level, which will involve detaching the carburettor concerned, inverting it and removing the float chamber bowl. If the float level is correct the distance between the uppermost portion of the floats and the flange of the mixing chamber body will be as follows:

T250	VM24SH carburettors	25.7 mm
T305	VM32SH carburettors	29.2 mm
T350	VM32SH carburettors	27.5 mm

Adjustments are made by bending the tang on the float arm in the direction required. (See accompanying diagram).

3 Replace the carburettor and turn the slow running screw until it is closed fully. Then turn it back approximately 1¾ turns. Adjust the amount of play in the throttle cable to within 0.5 - 1 mm and then start the engine.

4 Readjust the slow running screw until the engine runs smoothly at a stabilized engine speed. Note that a two-stroke engine will never fire evenly at very low engine speeds, so the setting will take the form of the best compromise. It will be necessary also to adjust the throttle stop screw in conjunction with the slow running screw, to achieve a stabilised engine speed. Fig. 2.1. shows the location of these adjusting screws.

5 Stop the engine, replace the sparking plug and lead and then repeat the procedure with the other cylinder and carburettor, after removing the sparking plug from the cylinder that has just been adjusted.

8 Synchronizing the carburettors

1 Power output will be unbalanced unless both carburettors work in perfect harmony with each other. Many cases of poor performance and low power output can be traced to carburettors that are out of phase with each other.

2 It is imperative to check that both carburettor slides enter the bore of the carburettor at the same time. Remove the air cleaner hoses and open the throttle wide (engine dead) so that

both slides are raised to their maximum height. Slowly close the throttle and check that both slides enter the carburettor bores at the same time. If they do not, make adjustments with either cable adjuster until they are in phase. Check that both slides close fully when the throttle is shut. Reconnect the air filter hoses and run the engine.

3　If carburettor adjustments are required, they should be made BEFORE the carburettors are synchronized. Always make the synchronizing adjustments with the engine warm, to obviate the risk of a false setting.

9 Carburettor settings

1　Some of the carburettor settings, such as the sizes of the needle jets, main jets and needle positions are predetermined by the manufacturer. Under normal circumstances it is unlikely that these settings will require modification, even though there is provision made. If a change appears necessary, it can often be traced to a developing engine fault.

2　As a rough guide, the slow running screw controls the engine speed up to 1/8th throttle. The throttle valve cutaway controls the engine speed from 1/8th to ¼ throttle and the position of the needle from ¼ to ¾ throttle. The main jet is responsible for the engine speed at the final ¾ to full throttle. It should be added that none of these demarkation lines is clearly defined; there is a certain amount of overlap between the carburettor components involved.

3　Always err on the side of a rich mixture because a weak mixture has a particularly adverse effect on the running of any two-stroke engine. A weak mixture will cause rapid overheating, which may eventually promote engine seizure. Reference to Chapter 3 will show how the condition of the sparking plugs can be used as a reliable guide to carburettor mixture strength.

10 Exhaust system - cleaning

1　The exhaust system is often the most neglected part of any two-stroke despite the fact that it has quite a pronounced effect on performance. It is essential that the exhaust system is inspected and cleaned out at regular intervals because the exhaust gases from a two-stroke engine have a particularly oily nature which will encourage the build-up of sludge. This will cause back pressures and affect the 'breather' of the engine.

2　Cleaning is made easy by fitting the silencers with detachable baffles, held in position by a set screw that passes through each silencer end. If the screw is withdrawn, the baffles can be drawn out of position for cleaning.

3　A wash with a petrol/paraffin mix will remove most of the oil and carbon deposits, but if the build-up is severe it is permissible to heat the baffles with a blow lamp and burn off the carbon and old oil.

4　At less frequent intervals, such as when the engine requires decarbonising, it is advisable also to clean out the exhaust pipes. This will prevent the gradual build-up of an internal coating of carbon and oil, over an extended period.

5　Do not run the machine with the baffles detached or with a quite different type of silencers fitted. The standard production silencers have been designed to give the best possible performance whilst subduing the exhaust note. Although a modified exhaust system may give the illusion of greater speed as a result of the changed exhaust note, the chances are that performance will have suffered accordingly.

6　When replacing the exhaust system, use new sealing rings at the exhaust port joints and check that the baffle retaining screws are tightened fully in the silencer ends.

11 Air cleaner - dismantling, servicing and reassembling

1　Before access can be gained to the air cleaner box and hoses it is necessary first to remove the left-hand cover that protects the battery and wiring.

2　To remove the air cleaner assembly, detach the two hoses from the carburettor intakes by slackening the retaining clips around the bell mouths. Remove also the two screws that secure the air cleaner assembly to the frame of the machine. The air cleaner will now lift away, complete with hoses.

3　The air cleaner element is contained within the air cleaner box. The box is separated by withdrawing the three cross head screws and washers that secure the lid to the main body of the assembly. The lid will pull away, with the air cleaner element held captive.

4　The resin-impregnated paper element is cleaned by using a blast of compressed air to blow away the accumulation of dust and other foreign matter. Clean both the inside and the outside.

5　If the element is damp or badly choked, it is advisable to replace the element. A dirty or blocked element will increase the rate of petrol consumption and will cause a marked fall-off in performance.

6　Do not run the machine with the air cleaner detached, even for a short period. The carburettors are jetted to compensate for the addition of the air cleaner and the carburation will be upset badly if it is removed. The effect will be overheating as a direct result of the permanently-weakened mixture strength and the risk of subsequent engine damage.

7　The air cleaner is reassembled by reversing the dismantling procedure. If the rubber hoses are split or perished, renew them in order to preclude the possibility of air leaks.

10.2. Screw in end of silencer releases baffle tube

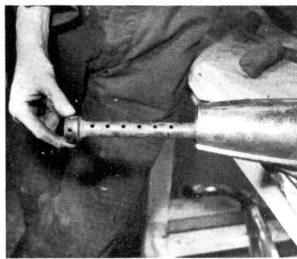

10.2a. Baffles withdraw completely for cleaning

11.4 Air cleaner box contains resin-impregnated paper element

12.1 Window in oil tank gives clear indication of contents. Acts as level indicator

12.5 Main feed to oil pump is bled by slackening large hexagonal nut

12.6 Individual oil lines are attached to pump by banjo unions

12.6a When banjo unions are disturbed, fit new sealing washers

12.6b Oil lines contain spring-loaded ball valve, to maintain pressure

12 Engine lubrication

1 The engine oil is contained in an oil tank mounted on the right-hand side of the machine. The tank holds 1.8 litres of oil and is fitted with an oil level inspection 'window' as a precaution against running out.

2 If the oil level falls below the screw in the centre of the inspection window, the tank must be replenished with oil of the correct viscosity. The machine should be standing on level ground when this check is made.

3 Oil flows from the oil tank to the oil pump by gravity, via a banjo union in the main outlet. The base of the outlet is fitted with a detachable cap that contains an integral magnet. This cap must be removed and cleaned during every three-monthly or 2,000 mile service.

4 The oil pump is driven by a worm and pinion arrangement from the primary drive of the engine. It is of the plunger type and must be replaced as a complete unit if it malfunctions.

5 The oil pump is interconnected to the throttle by means of a spring-loaded operating arm which is actuated by an extra control cable from the throttle cable junction box. Provision is included for bleeding the pump and also for adjusting the setting of the operating arm, as described in Chapter 1, Sections 40.2, 40.4. This procedure must be followed when the engine is dismantled or if the oil tank is permitted to run dry. Air bubbles will impede the free flow of oil.

6 In order to sustain a positive pressure, each of the oil pipe unions is fitted with a spring-loaded ball valve at the point where they connect with the crankcase drillings. If the ball valves are dismantled, it is essential that they are primed with oil prior to assembly and they are subsequently bled as detailed in Section 40.4 of Chapter 1.

7 A major advantage of the "Posi-Force" system is the fact that a pressure feed of oil will be maintained even when the throttle is closed. This obviates the risk of engine seizure that is liable to occur with petrol lubricated engines when the machine coasts down a long incline with the throttle closed.

Fault diagnosis - fuel system and lubrication

Symptom	Reason/s	Remedy
Engine gradually fades and stops	Fuel starvation	Check vent hole in filler cap and clear if blocked. Sediment in filter bowl or blocking float needle. Dismantle and clean.
Engine runs badly, black smoke from exhausts	Carburettors flooding	Dismantle and clean carburettors. Look for punctured float.
	Ruptured diaphragm in petrol tap	Remove feed pipe and check whether petrol flows when in normal running position. If so, replace diaphragm.
Engine lacks response and overheats	Weak mixture	Check for partial blockage in fuel/carburettors.
	Air cleaner disconnected or hoses split	Reconnect or repair.
White smoke from exhausts	Oil pump setting incorrect; too much oil passing	Check and reset oil pump.
	Incorrect oil in oil tank	Drain and refill with recommended grade
General lack of response to varying throttle openings	Blocked exhaust system	Remove silencer baffles and clean.

Chapter 3 Ignition system

Contents

Specifications

Sparking plugs

Make	NGK
Type(s)	B–77HC (standard)
	B–8H or B–9H (cooler)
	B–7H (hotter) B8ES (GT250 and Ram Air Models)
Reach	½ in (12.7 mm) ¾ in (GT250 and Ram Air Models)
Thread	14 mm
Gap	0.016 in - 0.020 in (0.4 - 0.5 mm)

Contact breakers

Gap	0.012 in - 0.016 in

Ignition timing	0.113 in (2.88 mm) BTDC both cylinders, measured by dial gauge
	24º BTDC both cylinders, measured by degree disc

1 General description

1 The spark necessary to ignite the petrol/air mix in the combustion chambers is derived from a battery and twin ignition coils. Each cylinder has its own separate contact breaker, condenser and coil to determine the exact moment at which the spark will occur. When the contact breaker points separate the low tension circuit is broken and a high tension voltage is developed by the coil which jumps the air gap across the points of the sparking plug and ignites the mixture.

2 An alternator attached to the left-hand end of the crankshaft assembly generates an alternating current which is rectified and used to charge the 12 volt, 5 amp/hr/battery. The rectifier is located below the nose of the dual seat, on the left-hand side of the machine, close to the battery. The twin ignition coils are mounted within the top frame tubes, close to the steering head of the machine. Both contact breakers and their condensers form part of the stator plate assembly that surrounds the generator rotor. The circuit is actuated by an ignition switch and key on the left-hand side of the machine, immediately below the nose of the petrol tank. This switch operates also the lights, both for running and parking.

2 Crankshaft alternator - checking the output

1 The output from the alternator mounted on the end of the

crankshaft can be checked only with specialised test equipment of the multi-meter type. It is unlikely that the average owner/rider will have access to this equipment or instruction in its use. In consequence, if the performance of the alternator is in any way suspect, it should be checked by a Suzuki agent or an auto-electrical specialist.

3 Ignition coils - checking

1 Each ignition coil is a sealed unit, designed to give long service without need for attention. They are located close together within the two top frame tubes, immediately to the rear of the steering head. If a weak spark and difficult starting cause the performance of either coil to be suspect, they should be tested by a Suzuki agent or an auto-electrical specialist who will have the appropriate test equipment. A faulty coil must be replaced; it is not possible to effect a satisfactory repair.

2 It is extremely unlikely that both coils will fail on the same occasion. If the complete ignition system fails, it is highly probable that the source of the fault will be found elsewhere. Apart from the common low tension supply, both coils work or. independent circuits.

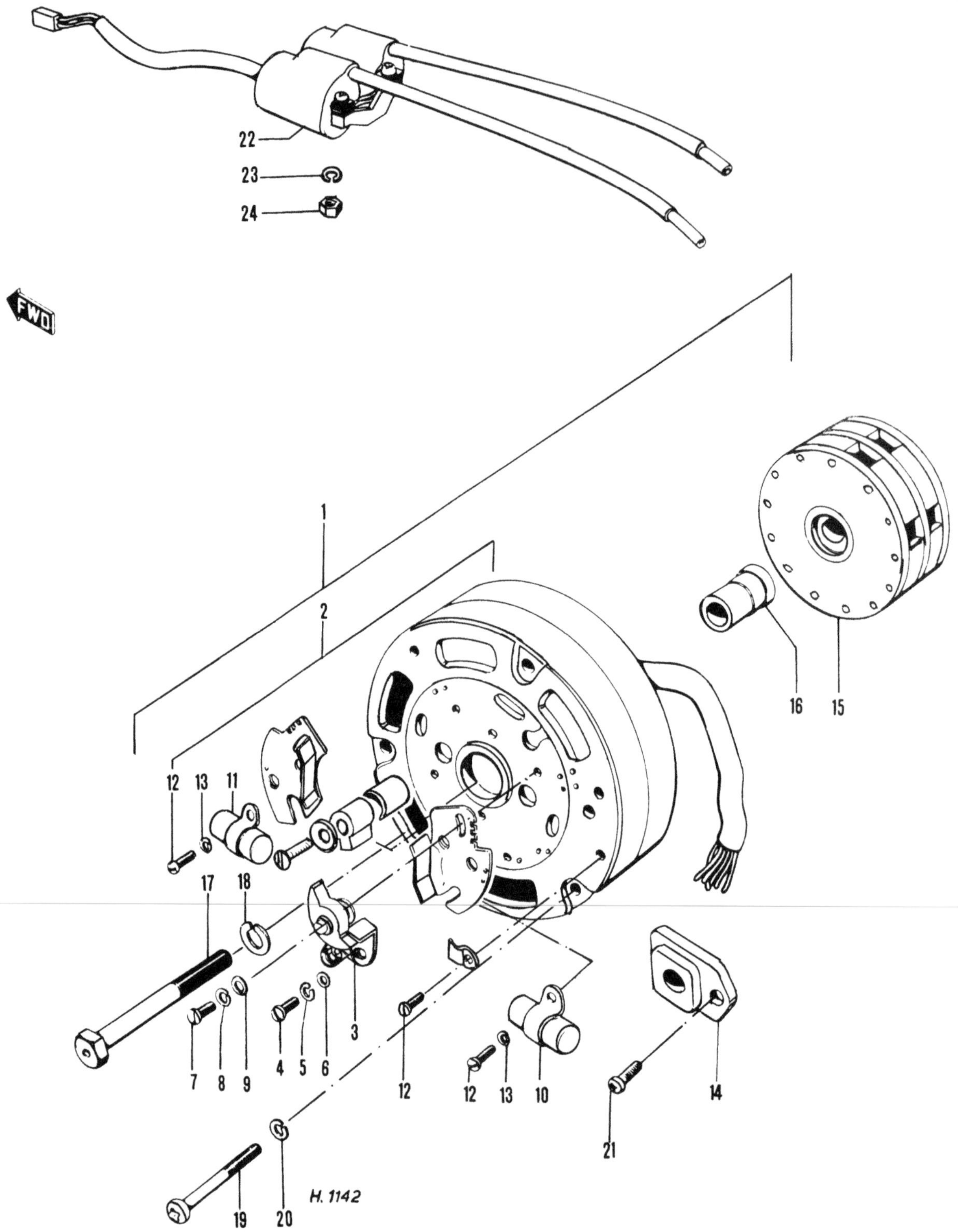

FWD

H. 1142

FIG.3.1. ALTERNATOR AND IGNITION COILS

1 Alternator assembly	7 Screw - 4 off	14 Wiring harness grommet	21 Screw
2 Stator assembly	8 Spring washer - 4 off	15 Alternator rotor	22 Ignition coil assembly
3 Contact breaker points	9 Flat washer - 4 off	16 Contact breaker cam	23 Spring washer 2 off
assembly - 2 off	10 Condenser - right hand	17 Rotor retaining bolt	24 Nut - 2 off
4 Screw - 2 off	11 Condenser - left hand	18 Spring washer	
5 Spring washer - 2 off	12 Screw - 3 off	19 Screw - 3 off	
6 Flat washer - 2 off	13 Spring washer - 2 off	20 Spring washer - 3 off	

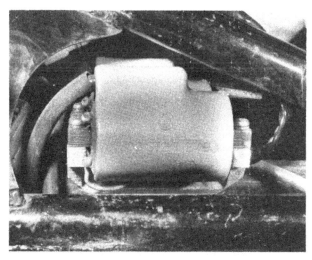

3.1. Ignition coils are mounted close to steering head

4 The fixed contact together with its mounting plate is removed by withdrawing the one retaining screw.

5 The points should be dressed with an oilstone or fine emery cloth. Keep them absolutely square during the dressing operation, otherwise they will make angular contact with each other when they are replaced and will quickly burn away.

6 Replace the contacts by reversing the dismantling procedure, making sure the insulating washers are replaced in the correct order. It is advantageous to apply a thin smear of grease to the moving contact pivot pin, prior to replacement of the contact arm.

7 Re-adjust the contact breaker gap before passing to the second contact breaker assembly. Repeat the dismantling, renovating, reassembly and readjustment procedure.

4 Contact breakers - adjustments

1 To gain access to the contact breaker assembly it is necessary to detach only the circular plate of the left-hand crankcase cover, which is retained in position by two cross head screws.

2 Rotate the engine slowly by means of the kickstarter until first one and then the other set of contact breaker points is in the fully-open position. Examine the faces of the contacts. If they are pitted or burnt it will be necessary to remove them for further attention, as described in Section 5 of this Chapter.

3 Adjustment is carried out by slackening the screw that retains the fixed contact plate in position and inserting the point of a screwdriver between the slot in the base plate and the projecting pegs, so that the contact can be moved in the direction desired. When the contact breaker gap is within the range 0.012 inch to 0.016 inch, as measured with a feeler gauge, tighten the clamp screw and recheck the gap. The feeler gauge must be a good sliding fit.

4 Repeat this adjustment for the second set of contact breaker points, making sure that the gap is EXACTLY identical with that of the other set of points.

5 Before replacing the cover plate, place a very light smear of grease on the contact breaker cam, making sure that none reaches the contact surfaces. The cover plate must be replaced with the drain hole downwards, to prevent water from entering the contact breaker housing.

5 Contact breaker points - removal, renovation and replacement

1 If the contact breaker points are burned, pitted or badly worn, they should be removed for dressing. If it is necessary to remove a substantial amount of material before the faces can be restored, the points should be replaced without question.

2 To remove the contact breaker points, slacken and remove the nut of the moving contact terminal to which the condenser lead is attached. Lift off the condenser lead and also the spring strip of the contact breaker arm, taking note of the way in which the insulating washers are assembled. Failure to observe the method of assembly may result in the contact being earthed inadvertently during reassembly, with the consequence isolation of the contact breaker circuit.

3 Detach the small circlip on the end of the moving contact pivot, using a pair of thin nose pliers. The moving contact complete with insulating arm, spring and washers can now be removed.

5.2. Contact breaker assembly with one set of points removed

5.3. Points assembly completely dismantled

5.7. Ensure points are fully open when checking gap

6 Condensers - removal and replacement

1 The condensers are included in the contact breaker circuitry to prevent arcing across the contact breaker points when they separate. Each condenser is connected in parallel with its own set of points and if a fault develops, ignition failure will occur in the circuit involved.
2 If the engine is difficult to start or if misfiring occurs on one cylinder, it is possible that the condenser in the ignition circuit of that cylinder has failed. To check, separate the contact breaker points by hand whilst the ignition is switched on. If a spark occurs across the points and they have a blackened and burnt appearance, the condenser can be regarded as unserviceable.
3 It is not possible to check the condenser without the appropriate test equipment. It is best to fit a replacement condenser in view of the low cost involved, and observe the effect on engine performance.
4 Each condenser is attached to the contact breaker base plate by a screw that passes through an integral clamp. It is necessary to remove also the condenser lead wire attachment to the terminal of the moving contact breaker arm.
5 It is highly improbable that both condensers will fail simultaneously. If this impression is given, commence looking for the source of the trouble in some other part of the ignition circuit.
6 Replace the new condenser by reversing the dismantling procedure. Make sure the clamp screw is tight because this forms the earth connection of the condenser. Check also that the insulating washers at the condenser lead terminal connection are in the correct order, to prevent electrical isolation and the reoccurrence of arcing across the points.

7 Ignition timing - checking and resetting

1 It cannot be overstressed that good performance is dependent on the accuracy with which the ignition timing is set. Even a small error can cause a marked reduction in performance and the possibility of engine damage, particularly to the pistons. Although the timing marks will verify whether the timing is accurate within certain limits, it is preferable to cross-check with a dial gauge and timing tester. Insert the dial gauge through the sparking plug hole in the left-hand cylinder head and set the ignition timing so that the points commence to separate when the piston is the recommended distance before top dead centre (see Specifications). The timing tester will show when the points have broken contact. Repeat this procedure for the right-hand cylinder.

2 For those who prefer to use a timing disc attached to the end of the crankshaft, the corresponding setting in degrees BTDC should be used (see Specifications).
3 If the ignition timing is correct, the inscribed lines on the rotor of the alternator will coincide EXACTLY with the line inscribed on the fixed stator plate as the contact breaker points for the cylinder concerned are on the point of separation.
4 The black line on the rotor applies to the timing of the right-hand cylinder (right-hand set of points) and the red line to the left-hand cylinder (left-hand set of points). The GT250A models have the old 'M' timing mark and the new 'A' timing mark stamped on the generator rotor. The 'A' marks, which are painted in red, are the ones to be used.
5 Before checking or resetting the ignition timing, always ensure the contact breaker gaps are correct. If the gaps are altered after the timing has been checked, some variation of the accuracy of the ignition timing will be inevitable.
6 If the timing is incorrect, the position of the contact breaker points in relation to the contact breaker cam can be adjusted within fixed limits. It is necessary to slacken the base plate screws and move the complete points assembly in the direction required, using a screwdriver between the slots at the extreme edge of the base plate and the projecting pegs close by. Re-tighten the base plate screws when the timing adjustment is correct, and check again.
7 The contact breaker circuit does not contain any provision for either advancing or retarding the ignition timing whilst the engine is running.

8 Sparking plugs - checking and resetting the gaps

1 A matched pair of 14 mm short-reach sparking plugs are fitted to each of the 250 cc and 350 cc twin cylinder models.
2 All models are fitted with an NGK type B-77HC sparking plug as standard, gapped within the range 0.016 inch to 0.020 inch (0.4 - 0.5 mm). The correct reach is ½ inch (12.7 mm). Certain operating conditions may ictate a change in sparking plug grade; if a colder plug is required, the NGK type B-8H or B-9H is recommended. An NGK type B-7H sparking plug is recommended for the other end of the scale, where a hotter plug is the requirement.
3 The following table shows the various plug recommendations with their British and Continental equivalents. Always use the grade of plug recommended to obviate the risk of ignition problems.
4 Check the gap of the plug points during every three-monthly or 2,000 mile service. The sparking plug has a particularly hard life in any high performance two-stroke engine. To reset the gap, bend the outer electrode to bring it closer to the centre electrode and check that a 0.016 inch feeler gauge can be inserted. Never bend the centre electrode or the insulator will crack, causing engine damage if particles fall in whilst the engine is running.
5 The condition of the sparking plug electrodes and insulator can be used as a reliable guide to engine operating conditions, with some experience. See the accompanying diagram.
6 Always carry a pair of sparking plugs of the correct grade. In the rare event of plug failure, they will enable the engine to be restarted.
7 Beware of over-tightening the sparking plugs, otherwise there is risk of stripping the threads from the aluminium alloy cylinder heads. The plugs should be sufficiently tight to seat firmly on their copper sealing washers and no more. Use a spanner that is a good fit, to prevent the spanner from slipping and breaking the insulator.
8 If the threads in the cylinder heads strip as a result of over-tightening the sparking plugs it is possible to reclaim the heads by the use of a "Helicoil" thread insert. This is a cheap and convenient way of replacing the threads; many motor cycle dealers operate a service of this nature specifically for 14 mm inserts.
9 Make sure the plug insulating caps are a good fit and have their rubber seals. These caps contain the suppressors that eliminate both radio and tv interference.

Electrode gap check - use a wire type gauge for best results

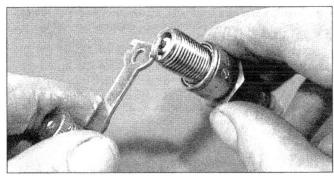

Electrode gap adjustment - bend the side electrode using the correct tool

Normal condition - A brown, tan or grey firing end indicates that the engine is in good condition and that the plug type is correct

Ash deposits - Light brown deposits encrusted on the electrodes and insulator, leading to misfire and hesitation. Caused by excessive amounts of oil in the combustion chamber or poor quality fuel/oil

Carbon fouling - Dry, black sooty deposits leading to misfire and weak spark. Caused by an over-rich fuel/air mixture, faulty choke operation or blocked air filter

Oil fouling - Wet oily deposits leading to misfire and weak spark. Caused by oil leakage past piston rings or valve guides (4-stroke engine), or excess lubricant (2-stroke engine)

Overheating - A blistered white insulator and glazed electrodes. Caused by ignition system fault, incorrect fuel, or cooling system fault

Worn plug - Worn electrodes will cause poor starting in damp or cold weather and will also waste fuel

Plug recommendations

Thread Size	Heat Range	NGK	Nippon-denso	Champion	AC	Auto-Lite	Bosch	Lodge
14 mm	Hot	B-6HS	W17F		44F	AE3		
X				L87	42F			
				L7, L7J	44FF		W225T1	H14, HN
12.7 mm (1/2'')		B-7HS	W22F	L5	42FF	AE2	W240T1	2HN, 3HN
Reach		B-77HC	W24F	L4J, L62R				
X					MC41F	AE23	W240P11S	
20.6 HEX		B-8HS	W24F		M42FF			
	Cold	B-9H		L4J			W260T1	HH14

Fault diagnosis - Ignition system

Symptom	Reason/s	Remedy
Engine will not start	Discharged battery Short circuit in wiring system	Recharge battery with a battery charger. Check wiring to locate source of fault.
Engine misfires and eventually stops	Whiskered sparking plugs Oiled sparking plugs	Replace plugs, using a hotter grade. Replace plugs, using a 'softer' grade.
Engine lacks response and overheats	Reduced contact braker gaps	Check and reset gaps.
Engine 'fades' when under heavy load	Pre-ignition	Replace plugs, using only recommended grades.

Chapter 4 Frame and forks

Contents

1 General description

The Suzuki 250 cc and 350 cc twin cylinder models utilise a common frame and fork assembly of conventional design. The front forks are of the telescopic type, with oil-filled one-way damper units. The frame is the full cradle type employing duplex tubes. Rear suspension is provided by a swinging arm fork, controlled by adjustable rear suspension units. These latter units are hydraulically-damped.

2 Front forks - removal from the frame

1 It is unlikely that the front forks will need to be removed from the frame as a complete unit unless the steering head bearings require attention or the forks are damaged in an accident.

2 Commence operations by removing either the control cables from the handlebar control levers or the levers complete with cables. The shape of the handlebars fitted and the length of the control cables will probably dictate which method is used.

3 Detach the handlebars together with their mountings. The mountings pass through rubber bushes in the top yoke of the forks, which act as a vibration damper. They are each retained by a castellated nut and washer which can be detached after the split pin has been removed.

4 It will be necessary to remove also the steering damper knob before the handlebar mountings can be lifted out of their rubber bushes. Remove the split pin through the extreme bottom end of the steering damper rod and unscrew the damper knob until it lifts clear complete with the rod attached. Detach also the headlamp complete which is held to the upper fork shrouds by two bolts.

5 If the machine is not already resting on the centre stand, support it in this fashion on firm, level ground. Balance it so that the front wheel is well clear of the ground.

6 Remove the speedometer drive cable from the gearbox in the front hub by unscrewing the round coupling nut. Detach also the front brake cable by unscrewing the adjusting nut and withdrawing the cable together with its outer cable adjuster and rubber gaiter from the brake hub anchorage. The wheel can now be released from the forks by unscrewing the clamp bolt through the bottom of the right hand fork leg and withdrawing the wheel spindle which has two flats on the end to accept a spanner. The spindle will unscrew from the left hand fork leg and pull clear.

7 Remove the front wheel complete with brake plate assembly. If desired, the front mudguard can be removed at this stage by detaching the nuts and bolts that retain the fork stays to the lugs on the lower fork legs and the two bolts on either side of the mudguard bridge that thread into the lower fork legs. Free both the speedometer cable and the brake cable from the guide attached to the lower left-hand end of the front mudguard and from the cable clips on the bottom yoke of the telescopic forks. These clamps are bent easily without disturbing the retaining bolts.

8 Unscrew the large diameter nut in the centre of the top yoke of the forks and remove it together with washer and clip that bears on the inside of the steering damper knob. Detach the speedometer drive cable and the tachometer drive cable from their respective indicators and remove both of the large bolts that thread into the top of each fork leg. The bolts will release an 'O' ring seal and a seating washer. Late models have an internal circlip to retain a cap, which can be removed at a later stage.

9 The top yoke of the forks can now be pulled clear together with the speedometer and tachometer heads, and the flashing indicator lamps.

10 Whilst the forks are supported in position, remove the steering head lock nut and washer. The lock nut is slotted, to facilitate the use of a 'C' spanner. The forks are now completely free and can be withdrawn from the bottom of the steering head stem. It may be necessary to raise the machine even higher so that they can be pulled clear of the frame completely.

11 Note that when the forks are removed from the frame, the uncaged ball bearings in the top and bottom steering head races are liable to fall free. It is advisable to wrap some rag around each race as it separates, to trap the bearings as they work free.

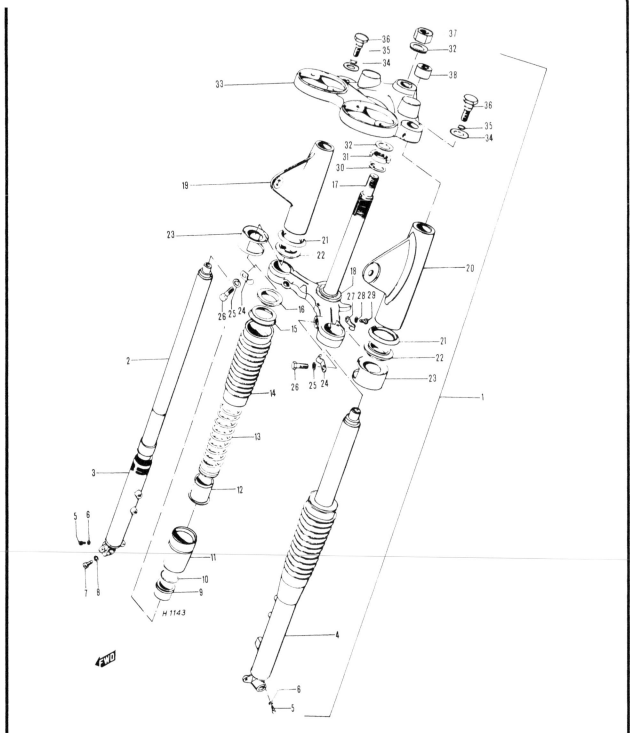

H 1143

FWD

FIG.4.1. FRONT FORKS

1   Front fork assembly complete
2   Fork inner tube - 2 off
3   Right hand outer tube
4   Left hand outer tube
5   Drain plug - 2 off
6   Drain plug gasket - 2 off
7   Pinch bolt
8   Washer
9   Inner tube guide bush - 2 off
10 'O' ring seal - 2 off
11 Outer tube threaded collar - 2 off
12 Dust seal - 2 off
13 Fork spring - 2 off
14 Rubber gaiter - 2 off
15 Spring guide - 2 off
16 Spring seat - 2 off
17 Steering head stem
18 Steering head cup - 2 off
19 Right hand upper fork shroud
20 Left hand upper fork shroud
21 Shroud seat - 2 off
22 Shroud cushion - 2 off
23 Shroud bracket - 2 off
24 Speedometer cable clip - 2 off
25 Spring washer - 2 off
26 Bottom yoke clamp bolt - 2 off
27 Brake cable clip
28 Spring washer
29 Screw
30 Steering head stem lock washer
31 Steering head stem locknut
32 Steering head stem washer
33 Fork top yoke
34 Flat washer - 2 off
35 'O' ring seal - 2 off
36 Fork top bolt - 2 off
37 Top yoke retaining nut
38 Bush for top yoke

2.6 Detach speedometer drive cable from front wheel brake plate

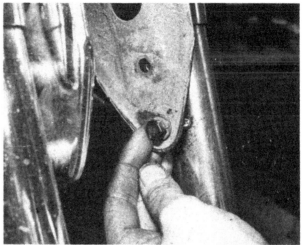
2.7b...... also by bridge pieces bolted to inside of legs

2.7. Remove front wheel complete with brake plate

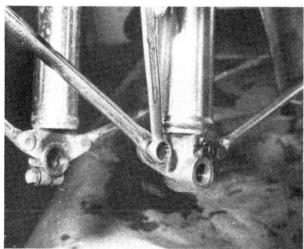
2.7a. Mudguard is attached by stays to lower ends of fork legs......

3 Front forks - dismantling

1 It is advisable to dismantle each fork leg separately, using an identical procedure. There is less chance of unwittingly interchanging parts if this approach is adopted. Commence by draining the forks; there is a drain screw in each leg, above the wheel spindle.

2 The chromium plated collar that acts as the upper anchorage for the rubber gaiter will remain captive to the lower fork yoke. Slacken the pinch bolt through the lower fork yoke and withdraw the complete fork leg together with the spring, rubber gaiter and the spring seating.

3 Unscrew the chromium plated collar that threads on to the lower outer leg of the forks. This collar has a normal right-hand thread and should not prove too difficult to unscrew if the fork leg is held in a vice. A wide rubber band around the collar will improve the grip; in an extreme case it may be necessary to use a strap or a chain spanner.

4 Remove the collar and withdraw the inner fork tube complete with limit ring and lower bush. The top bush and 'O' ring seal will be a sliding fit over the inner fork tube after they have been released from the top of the lower fork leg by the screwed collar. The fork leg is now dismantled completely, leaving the headlamp and top fork shrouds in position. Repeat this procedure for the other fork leg.

5 If it is desired to dismantle either or both fork legs without disturbing the steering head races, follow the procedure given in the preceding Section from paragraphs 5 to 7 inclusive. Then continue from paragraph 2 of this Section, after first removing the large chromium plated bolt that threads into the top of the fork leg, through the top yoke of the forks.

4 Steering head bearings - examination and renovation

1 Before commencing reassembly of the forks, examine the steering head races. The ball bearing tracks of the respective cup and cone bearings should be polished and free from indentations or cracks. If signs of wear or damage are evident, the cups and cones must be replaced. They are a tight push fit and should be drifted out of position.

2 Ball bearings are cheap. If the originals are marked or discoloured, they should be replaced without question. To hold the steel balls in position during reattachment of the forks, pack the bearings with grease. Note that each race should have space to include one extra ball. This space is necessary to prevent the ball bearings from skidding on each other, which would accelerate the rate of wear.

3.2. Pinch bolt in lower fork yoke must be slackened to release leg

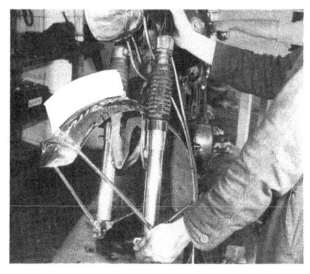

3.2a. Leg will pull free when top bolt is withdrawn

3.2b. Lift off gaiter and fork spring before collar is unscrewed

3.2c. Spring seats in screwed collar, over dust seal

3.3. Collar unscrews from fork leg (right-hand thread)

3.4. Top bush is sliding fit over inner fork tube

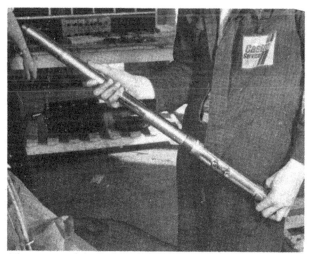

5.3. Lower bush is permanently attached to lower end of inner fork tube

5.4. Main oil seal is integral part of screwed collar

6.1. Front brake plate must locate with projection on left-hand fork leg

5 Front forks - examination and renovation

1 The parts most liable to wear over an extended period of service are the bushes that fit over the inner tubes of each fork leg and the oil seals contained within each screwed collar. Worn fork bushes normally cause judder when the front brake is applied and the increased amount of play can be detected by pulling and pushing on the handlebars when the front brake is full on.

2 Replacement of the worn bushes is quite straightforward. The upper bush fits under the screwed collar of the lower outer fork leg and acts mainly as a guide for the fork inner tube. It has an 'O' ring seal between its upper lip and the screwed collar, to prevent oil leakage.

3 The lower bush is permanently attached to the bottom end of the fork inner tube by means of a peg. Specialist attention is required when this bush wears and it is recommended that the replacement is fitted by a Suzuki agent.

4 The main oil seal forms an integral part of the screwed collar and if the seal becomes worn or damaged, the collar complete must be renewed.

5 Check the fork springs for wear. If wear is evident, or if the springs have taken a permanent set, they must be replaced, always as a pair.

6 Check the outer surface of the fork inner tubes for scratches or roughness. It is only too easy to damage the main oil seal during reassembly, if this precaution is not observed.

7 Inspect also the dust seals, on which the bottom of each fork spring seats. If the seals are damaged, they will permit the ingress of foreign matter, which will impair fork efficiency.

8 It is rarely possible to straighten forks that have been bent as the result of accident damage, especially if the correct jigs are not available. It is always advisable to err on the safe side and fit new replacements, especially since there is no means of checking to what extent the damaged forks have been over-stressed.

9 Damping is effected by oil within each fork leg passing either in or out of small holes drilled in the fork leg itself. The damping action can be altered by changing the viscosity of the oil, although a change is unlikely to prove necessary in most cases.

6 Front forks - replacement

1 Replace the front forks by reversing either of the dismantling procedures described in Sections 2 and 3 of this Chapter, whichever is the more appropriate. Make sure brake plate locates with projection on fork leg.

2 Before fully tightening the front wheel spindle, fork yoke pinch bolts and the bolts in the top of each fork leg, bounce the forks several times to ensure they work freely and are clamped in their original settings. Complete the final tightening from the front wheel spindle upwards.

3 Do not forget to add the damping fluid to each fork leg before the bolts in the top of each fork leg are finally tightened. Each fork leg should be filled with the following amount of SAE 30 oil. Check that the drain plugs have been replaced in the bottom of each fork leg before the oil is added!

T250J	190 ccs		
T250/II	220 ccs	T350R and J	220 ccs
T250R	180 ccs		

4 Difficulty is often experienced when attempting to draw the fork inner tube into position in the top yoke, during reassembly. It is worth while making up a special tool to aid the correct location, which takes the form of a threaded rod of the correct diameter, to which a 'T' handle is attached. See the accompanying drawing.

5 Check the adjustment of the steering head bearings before the machine is used on the road and again shortly afterwards. If the bearings are too slack, fork judder will occur. There should be no play at the head races when the handlebars are pushed and pulled with the front brake applied hard.

6 Overtight head races are equally undesirable. It is possible to place a pressure of several tons on the head bearings by over-tightening, even though the handlebars seem to turn freely

Overtight head bearings will cause the machine to roll at low speeds and give imprecise steering. Adjustment is correct if there is no play in the bearings and the handlebars swing to full lock either side when the machine is on the centre stand with the front wheel clear of the ground. Only a light tap on each end should cause the handlebars to swing.

7 Steering head lock

1 The steering head lock is attached to the upper surface of the lower fork yoke by two screws. When in the locked position, a plunger is lowered, which falls between two projections cast in the yoke when the handlebars are on full left-hand lock. The handlebars cannot be straightened until the lock is released and the plunger raised.

2 If the lock malfunctions, it must be replaced. A repair is impracticable. When the lock is changed it follows that the key must be changed too, to match the new lock.

8 Steering damper - function and use

1 A steering damper is a fitting that provides means of adding friction to the steering head assembly so that the front forks will turn less easily. It is a relic of the early days of motor cycling when machines were liable to develop 'speed wobbles' which grew in intensity until the rider was unseated. With today's more sophisticated front fork damping and improved frame designs, a steering damper is almost a superfluous fitting unless a sidecar is attached to the machine or very poor road surfaces are encountered.

2 The steering damper is, in effect, a small clutch without any compression springs. When the steering damper knob is tightened, the friction discs and the plain discs are brought into closer proximity with each other and it is more difficult to deflect the handlebars from their set position. If the knob is tightened fully, the handlebars are virtually locked in position. Under normal riding conditions, the steering damper should be slackened off. Only at very high speeds or on rough surfaces is there any need to apply some damper friction.

3 The steering damper assembly will be found at the base of the steering head column, immediately below the bottom fork yoke. The centre fixed plate is attached to the yoke by the left-hand lock stop.

4 Although it is unlikely that the steering damper assembly will require attention during the normal service life of the machine, it can be removed by withdrawing the damper knob and threaded rod, following the procedure given in Section 2.4 of this Chapter. The remainder of the assembly is freed when the left-hand lock stop of the bottom fork yoke is unscrewed.

9 Frame - examination and renovation

1 The frame is unlikely to require attention unless accident damage has occurred. In some cases, replacement of the frame is the only satisfactory course of action if it is out of alignment. Only a few frame repair specialists will have the necessary jigs and mandrels essential for re-setting the frame and even then there is no means of assessing to what extent the frame may have been overstressed.

2 After a machine has covered an extensive mileage, it is advisable to inspect the frame for signs of cracking or splitting at any of the welded joints. Rust corrosion can also cause weaknesses at these joints. Minor repairs can be effected by welding or brazing, depending on the extent of the damage.

3 Remember that a frame that is out of alignment will cause handling problems and may even promote the 'speed wobbles'. If misalignment is suspected, as the result of an accident, it will be necessary to strip the machine completely so that the frame can be checked and if needs be, replaced.

10 Swinging arm rear fork - dismantling, examination and renovation

1 The rear fork of the frame assembly pivots on a detachable bush within each end of the fork cross member and a pivot which passes through frame lugs and the centre of each of the two bushes. It is quite easy to renovate the swinging arm pivots when wear necessitates attention.

2 To remove the swinging arm fork, first detach the rear chainguard and the final drive chain. The former is retained by a countersunk cross head screw in front of the mounting for the bottom of the left-hand rear suspension unit and by a bolt through a lug which locates with the inner portion of the fork cross member, on the left-hand end of the radiused section. It is best to separate the chain whilst the spring link is resting in the rear wheel sprocket.

3 Detach the rear brake torque arm from the brake plate by removing the spring clip and withdrawing the securing nut, washer and bolt. Remove also the cable from the brake plate. Then withdraw the rear brake cable by unscrewing the adjuster and freeing the rear wheel complete with sprocket and brake drum by unscrewing the inner of the two left-hand spindle nuts, withdrawing the wheel spindle, then removing the large diameter left-hand nut. Do not lose the distance piece which will fall clear from the right-hand side of the wheel, as it is removed.

4 Remove both rear suspension units. Each is retained in position by a domed nut and a washer. When the nuts and washers have been removed, the units can be pulled off their mounting studs.

5 Remove the self-locking nut from the right-hand end of the swinging arm pivot pin. Withdraw the pin and pull the swinging arm fork from the frame.

6 If the dust cap is removed from each end of the fork cross member, the 'O' ring seal, thrust washer and the pivot bearing bush can be withdrawn. Three separate spacers are carried between the two bearing bushes; these need not be removed because they play no part in the bearing system apart from acting as a general support and guide for the pivot pin.

7 Wear will take place in the bearing bushes, both of which should be replaced. Replace both the bearing bushes and the pivot pin if the clearance between them exceeds 0.014 inch (0.35 mm) or the pivot pin if it is out of true by more than 0.020 inch (0.5 mm).

8 Reassemble the swinging arm fork by reversing the dismantling procedure. Grease the pivot pin and both bearing bushes liberally prior to reassembly and check that the grease nipple fitted in the top cross member of the fork is not obstructed.

9 Worn swinging arm pivot bearings will give imprecise handling with a tendency for the rear of the machine to twitch or hop. The play can be detected by placing the machine on its centre stand and with the rear wheel clear of the ground pushing and pulling sideways on the fork ends.

11 Rear suspension units - examination

1 Rear suspension units of the hydraulically-damped type are fitted to all the Suzuki 250 cc and 350 cc twins. They can be adjusted to give three different spring settings, without removal from the machine.

2 Each rear suspension unit has two peg holes immediately above the adjusting notches, to facilitate adjustment. Either a 'C' spanner or a metal rod can be used to turn the adjusters. Turn clockwise to increase the spring tension and stiffen up the suspension. The recommended settings are as follows:

Position 1 (least tension)	Normal running without a pillion passenger.
Position 2 (middle setting)	High speed touring

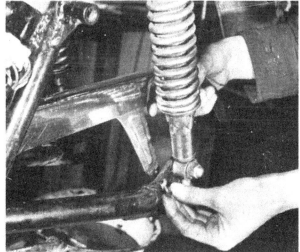

10.2. Countersunk screw retains rear end of chainguard

10.5a Pull swinging arm fork from frame

10.4. Rear suspension units are retained on studs by dome nuts

10.6. Dust caps fit over ends of fork pivot......

10.5. Withdraw pivot pin and ......

10.6a......contain 'O' ring seals and.......

10.6b thrust washers

10.7. Bushes and pivot pin must be replaced to eliminate wear

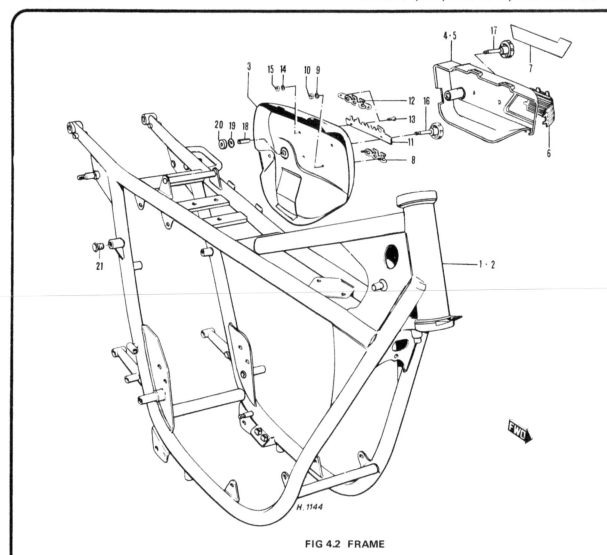

H. 1144

FIG 4.2 FRAME

1-2 Frame assembly	8 Capacity identification	identification plate	18 Cover knob spacer
3-5 Left hand frame cover	plate	13 Screw - 2 off	19 Cover knob nut
alternatives	9 Spring washer 2 off	14 Spring washer - 2 off	20 Cover knob washer
6 Frame cover trim	10 Nut - 2 off	15 Nut - 2 off	21 Blanking off cap -
7 Identification strip	11 Manufacturer's name plate	16-17 Frame cover knob	2 off
(self adhesive)	12 Alternative capacity	alternatives	

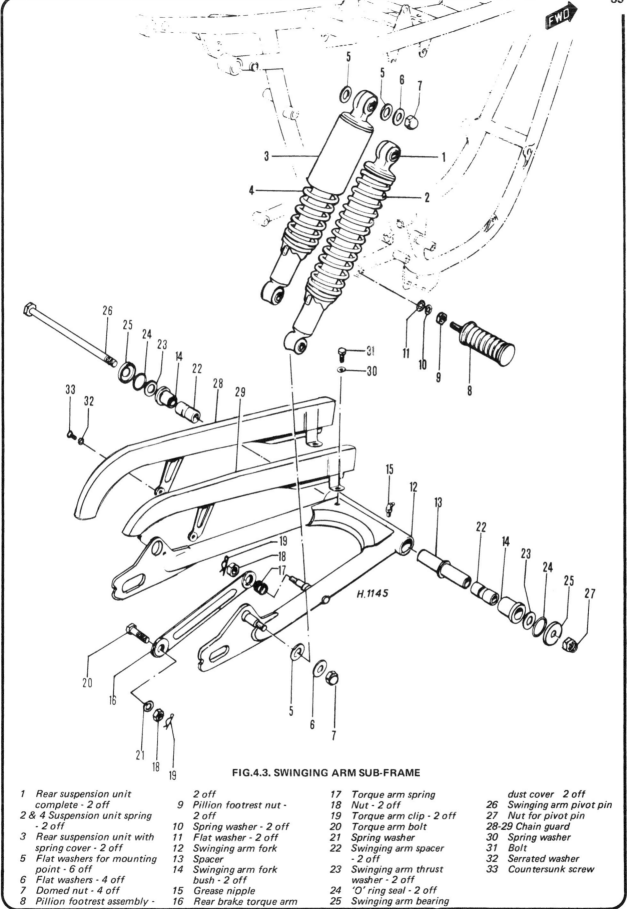

FIG.4.3. SWINGING ARM SUB-FRAME

H.1145

1	Rear suspension unit complete - *2 off*		
2 & 4	Suspension unit spring - *2 off*		
3	Rear suspension unit with spring cover - *2 off*		
5	Flat washers for mounting point - *6 off*		
6	Flat washers - *4 off*		
7	Domed nut - *4 off*		
8	Pillion footrest assembly -		

1   Rear suspension unit complete - *2 off*
2 & 4 Suspension unit spring - *2 off*
3   Rear suspension unit with spring cover - *2 off*
5   Flat washers for mounting point - *6 off*
6   Flat washers - *4 off*
7   Domed nut - *4 off*
8   Pillion footrest assembly -

    2 off
9   Pillion footrest nut - *2 off*
10 Spring washer - *2 off*
11 Flat washer - *2 off*
12 Swinging arm fork
13 Spacer
14 Swinging arm fork bush - *2 off*
15 Grease nipple
16 Rear brake torque arm

17 Torque arm spring
18 Nut - *2 off*
19 Torque arm clip - *2 off*
20 Torque arm bolt
21 Spring washer
22 Swinging arm spacer - *2 off*
23 Swinging arm thrust washer - *2 off*
24 'O' ring seal - *2 off*
25 Swinging arm bearing

    dust cover *2 off*
26 Swinging arm pivot pin
27 Nut for pivot pin
28-29 Chain guard
30 Spring washer
31 Bolt
32 Serrated washer
33 Countersunk screw

Position 3 (highest tension)   High speed competition events or with pillion passenger and/or heavy loads.

3 There is no means of draining or topping up. If the damping of the suspension units fails, the units complete must be replaced.

4 In the interests of good roadholding it is essential that both suspension units have the same load setting.

12 Centre stand - examination

1 The centre stand is attached to lugs welded to the bottom frame tubes, and pivots on two bushes, one through each leg, which are retained by nuts and bolts. An extension spring is used to keep the stand in the fully-retracted position when the machine is in use.

2 Check that the return spring is in good condition and that both nuts and bolts at the pivots are tight. If the stand falls whilst the machine is in motion it may catch in some obstacle and unseat the rider.

13 Prop stand - examination

1 A prop stand is also fitted, when it is not desired to use the centre stand. The prop stand pivots from a metal plate attached to a lug on the lower left-hand frame tube by two bolts and is fitted with an extension spring to ensure the stand is retracted automatically immediately the weight of the machine is taken from it.

2 Check that the two bolts retaining the metal plate are both fully tightened and also the single pivot bolt through the eye of the stand arm. Check also that the extension spring is in good condition and is not over-stretched. An accident is almost inevitable if the stands extend whilst the machine is on the move!

14 Footrests - examination and renovation

1 Each footrest is bolted to a boss on the plate welded to each of the duplex frame tubes. They are non-adjustable for height and are prevented from turning by a second bolt which passes through an extension of the footrest arm into the tube that extends from each frame tube.

2 If the machine is dropped, it is probable that the rearmost bolt will shear and it will be necessary to drill out the stump after the footrest itself has been removed. If the footrest arm is bent, it can be straightened in a vice, using a blow lamp to warm the area where the bend occurs. The footrest rubber must be removed before any heat is applied.

15 Rear brake pedal - examination and renovation

1 The rear brake pedal pivots around a boss attached to the right-hand frame tube. It is held captive by the bolt that passes through the right-hand footrest, into the centre of the pivot boss. A coil spring around the pivot aids the return of the pedal to its normal operating position.

2 If the brake pedal is bent or twisted in an accident, it should be removed and straightened in a manner similar to that recommended for the footrests in the preceding Section.

16 Dual seat - removal and replacement

1 The dual seat is attached to the rear of the top frame tubes by a bracket with slotted ends. The nose of the seat locates with a tube across the top tubes of the frame, that is raised on two lugs.

2 To remove the dual seat, slacken each of the two nuts at the end bracket and withdraw the seat from the rear. It is replaced by re-engaging the nose with the cross tube and the slotted bracket with the nuts at the rear of the top frame tubes, which should then be tightened fully.

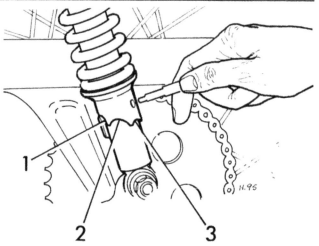

FIG.4.4. SETTINGS FOR REAR SUSPENSION UNITS

1 Normal riding   2 High speed   3 Pillion riding

16.1. Nose of dual seat locates with tube across frame tubes

17 Speedometer and tachometer heads - removal and replacement

1 The speedometer and tachometer heads fit within the alloy casting of the top yoke of the forks, or in the case of the earlier models, to a mounting bracket bolted to the top yoke. In either instance they are attached to the yoke or bracket by nuts and washers that thread on to studs that project from the base of each outer case.

2 Before either head can be removed, it is necessary first to detach the drive cables. Unscrew the circular coupling nut from the underside of each head and pull away the cables.

3 It will be necessary also to detach the various internal bulbs, which are mounted in rubber-covered bulb holders that push into the base of each instrument.

4 The heads can now be pulled clear from their mountings. The earlier instruments have a rubber cushioning mat to dampen out the effects of vibration. These also will be freed and should not be lost.

5 Apart from defects in either the drive or the drive cable, a speedometer or tachometer that malfunctions is difficult to repair. Fit a replacement, or alternatively entrust the repair to a competent instrument repair specialist.

6 Remember that a speedometer in correct working order is a statutory requirement in the UK. Apart from this legal requirement, reference to the odometer reading is the best means of keeping pace with the maintenance schedules.

18 Speedometer and tachometer drive cables - examination and maintenance

1   It is advisable to detach both cables from time to time in order to check whether they are lubricated adequately, and whether the outer coverings are compressed or damaged at any point along their run. Jerky or sluggish movements can often be attributed to a cable fault.

2   For greasing, withdraw the inner cable. After removing the old grease, clean with a petrol-soaked rag and examine the cable for broken strands or other damage.

3   Regrease the cable with high melting point grease, taking care not to grease the last six inches at the point where the cable enters the instrument head. If this precaution is not observed, grease will work into the head and immobilise the movement.

4   If either instrument head stops working, suspect a broken drive cable. Inspection will show whether the inner cable has broken; if so, the inner cable alone can be renewed and re-inserted in the outer casing, after greasing. Never fit a new inner cable alone if the outer covering is damaged or compressed at any point along its run.

18.4. Tachometer drive is taken from extension of oil pump drive

19 Speedometer and tachometer drives - location and examination

1   The speedometer drive gearbox is an integral part of the front wheel brake plate, and is driven internally from the wheel hub. The gearbox rarely gives trouble if it is lubricated regularly. If wear in the drive mechanism occurs, the worm can be withdrawn, complete with shaft, from the brake plate housing. The drive pinion that mates with the worm is secured within the brake plate also, by a circlip in front of the shaped driving plate that engages with slots in the front wheel hub.

2   The tachometer drive is taken from the oil pump, which shares a common drive facility. It is unlikely that the tachometer drive will give trouble.

20 Cleaning the machine

1   After removing all surface dirt with a rag or sponge that is washed frequently in clean water, the machine should be allowed to dry thoroughly. Application of car polish or wax to the cycle parts will give a good finish, particularly if the machine has not been neglected for a long period.

2   The plated parts of the machine should require only a wipe with a damp rag. If the plated parts are badly corroded, as may occur during winter when the roads are salted, it is permissible to use one of the proprietary chrome cleaners. These often have an oily base, which will help to prevent corrosion reoccurring.

3   If the engine parts are particularly oily, use a cleaning compound such as "Gunk" or "Jizer' Apply the compound whilst the parts are dry and work it in with a brush so that it has the opportunity to penetrate the film of grease and oil. Finish off by washing down liberally, taking care that water does not enter either the carburettors or the electrics. If desired, the now clean, polished aluminium alloy parts can be enhanced further by using a special polish such as Solvol "Autosol" which will restore them to full lustre.

4   If possible, the machine should be wiped over immediately after it has been used in the wet, so that it is not garaged in damp conditions that will promote rusting. Make sure to wipe the chain and re-oil it, to prevent water from entering the rollers and causing harshness with an accompanying rapid rate of wear. Remember there is little chance of water entering the control cables and causing stiffness of operation, if they are lubricated regularly as recommended in the Routine Maintenance Section.

Fault diagnosis - frame and forks

Symptom	Reason/s	Remedy
Machine veers either to the right or the left with hands off handlebars	Incorrect wheel alignment Bent frame Twisted forks	Check and realign. Check, and if necessary replace. Check, and if necessary replace.
Machine rolls at low speeds	Overtight steering head bearings	Slacken bearings.
Machine judders when front brake is applied	Slack steering head bearings	Tighten, until all free play is lost.
Machine pitches badly on uneven surfaces	Ineffective front fork dampers Ineffective rear suspension units	Check oil content of forks. Check whether units still have damping action.
Fork action stiff	Fork legs out of alignment (twisted in yokes)	Slacken yoke clamps, front wheel spindle and fork top bolts. Pump forks several times, then retighten from bottom upwards.
Machine wanders. Steering imprecise, rear wheel tends to hop	Worn swinging arm pivot	Dismantle and replace bushes and pivot pin.

Torque wrench settings	T250	T305	T350
Front fork bolts	25.3	25.3	25.3
Fork yoke bolts	57.8	57.8	57.8
Handlebar bracket bolts	10.8	10.8	10.8
Rear suspension unit nuts	18.1	18.1	18.1
Swinging arm pivot nut	43.4	43.4	43.4

Chapter 5 Wheels, brakes and tyres

Contents

Specifications

Tyres

	T250	GT250	T305	T350
Front	2.75 x 18 in.	3.00 x 18 in.	3.00 x 18 in.	3.00 x 18 in.
Rear	3.00 x 18 in.	3.25 x 18 in.	3.25 x 18 in.	3.25 x 18 in.

Brakes

Front	Twin leading shoe drum brake*
Rear	Conventional drum brake

*GT250 models have an hydraulically-operated disc brake

Chain

Number of rollers 96

Note: conversion kits are available for converting to British chain sizes. It is necessary to change the chain and both sprockets.

Torque wrench settings

	T250 and GT250	T305	T350
Front and rear wheel spindle	47.0	47.0	47.0

1 General description

1 Both wheels are of 18 inch diameter. They carry a ribbed tread tyre of 3.00 inch section on the front wheel and a block tread tyre of 3.25 inch section on the rear wheel. An exception occurs in the case of the early 250cc models, which have slightly smaller section tyres. All models employ steel wheel rims in conjunction with cast aluminium alloy hubs. Each wheel has, as standard an 8 inch diameter internal expanding brake, which is of the twin leading shoe variety in the case of the front wheel. The GT250 models are the exception to the rule; these models have an hydraulically-operated disc brake fitted to the front wheel, in place of the standard drum brake.

2 Both wheels are quickly detachable; the rear wheel can be removed from the frame without disturbing either the rear wheel sprocket or the final drive chain.

2 Front wheel - examination and renovation

1 Place the machine on the centre stand so that the front wheel is raised clear of the ground. Spin the wheel and check the rim alignment. Small irregularities can be corrected by tightening the spokes in the affected area, although a certain amount of experience is advisable to prevent over-correction. Any flats in the wheel rim should be evident at the same time. These are more difficult to remove and in most cases it will be necessary to have the wheel rebuilt on a new rim. Apart from the effect on stability, a flat will expose the tyre bead and walls to greater risk of damage if the machine is run with a deformed wheel.

2 Check for loose and broken spokes. Tapping the spokes is the best guide to tension. A loose spoke will produce a quite different sound and should be tightened by turning the nipple in an anti-clockwise direction. Always re-check for run-out by spinning the wheel again. If the spokes have to be tightened an excessive amount, it is advisable to remove the tyre and tube by following the procedure detailed in Section 17 of this Chapter. This is so that the protruding ends of the spokes can be ground off, to prevent them from chafing the inner tube and causing punctures.

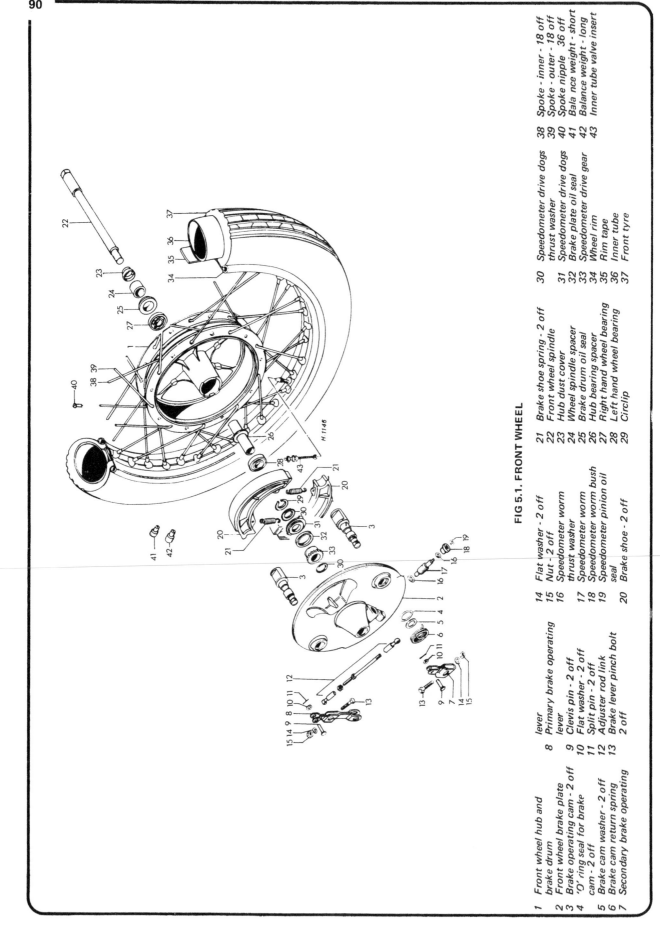

FIG 5.1. FRONT WHEEL

1 Front wheel hub and brake drum
2 Front wheel brake plate
3 Brake operating cam - 2 off
4 'O' ring seal for brake cam - 2 off
5 Brake cam washer - 2 off
6 Brake cam return spring
7 Secondary brake operating lever
8 Primary brake operating lever
9 Clevis pin - 2 off
10 Flat washer - 2 off
11 Split pin - 2 off
12 Adjuster rod link
13 Brake lever pinch bolt 2 off
14 Flat washer - 2 off
15 Nut - 2 off
16 Speedometer worm thrust washer
17 Speedometer worm
18 Speedometer worm bush
19 Speedometer pinion oil seal
20 Brake shoe - 2 off
21 Brake shoe spring - 2 off
22 Front wheel spindle
23 Hub dust cover
24 Wheel spindle spacer
25 Brake drum oil seal
26 Hub bearing spacer
27 Right hand wheel bearing
28 Left hand wheel bearing
29 Circlip
30 Speedometer drive dogs thrust washer
31 Speedometer drive dogs
32 Brake plate oil seal
33 Speedometer drive gear
34 Wheel rim
35 Rim tape
36 Inner tube
37 Front tyre
38 Spoke - inner - 18 off
39 Spoke - outer - 18 off
40 Spoke nipple - 36 off
41 Balance weight - short
42 Balance weight - long
43 Inner tube valve insert

3.1. Front brake plate assembly withdraws from hub

3.2. Examine the condition of the brake linings

3 4. Twin leading-shoe brake assembly has two operating cams

3 Front brake assembly - examination, renovation and re-assembly

All models except GT250

1 The front brake assembly complete with brake plate can be withdrawn from the front wheel hub after the wheel spindle has been pulled out and the wheel removed from the forks. Refer to Chapter 4, Section 2.6 for the correct procedure.

2 Examine the condition of the brake linings. If they are wearing thin or unevenly the brake shoes should be replaced. The linings are bonded on and cannot be supplied separately.

3 To remove the brake shoes, turn the brake operating lever so that the brake is in the full on position. Pull the brake shoes apart to free them from the operating cams after releasing the pivot pins from the fixed ends, and lift them away complete with their return springs by reverting to a 'v' formation. When they are well clear of the brake plate, the return springs can be removed and the shoes separated.

4 Before replacing the brake shoes, check that both brake operating cams are working smoothly and not binding in their pivots. The cams are removed for greasing by detaching their operating arm from the splined end of each shaft, after first slackening the pinch bolts. Before the arms are pulled off their respective shafts, mark their position on the splines to aid correct location. Do not alter the setting of the screwed rod that joins both arms, otherwise the brake setting will require re-adjustment after reassembly.

5 Check the inner surface of the brake drum, on which the brake shoes bear. The surface should be smooth and free from score marks or indentations, otherwise reduced braking efficiency will be inevitable. Remove all traces of brake lining dust and wipe with a rag soaked in petrol to remove all traces of grease or oil.

6 To reassemble the brake shoes on the brake plate, for the return spring and pull shoes apart, holding them in a 'v' formation. If they are now located with the brake operating cams and pivots they can be pushed back into position by pressing downwards. Do not use excessive force, or there is risk of distorting the shoes permanently.

4 Front wheel disc brake - examination and renovation

GT250 models

1 Check the front brake master cylinder, hose and caliper unit for signs of fluid leakage. Pay particular attention to the condition of the hose, which should be replaced without question if there are signs of cracking, splitting or other exterior damage.

2 Check also the level of hydraulic fluid by removing the cap on the brake fluid reservoir, diaphragm plate and diaphragm. This is one of the regular maintenance tasks, which should never be neglected. If the level is below the level mark, fluid of the correct grade must be added. NEVER USE ENGINE OIL or anything other than the recommended fluid. Other fluids have unsatisfactory characteristics and will rapidly destroy the seals.

3 The brake pads should also be inspected for wear. Each has a red line around its outer edge, which denotes the limit of wear. When this limit has been reached, BOTH pads must be replaced, even if only one has reached the limit line. Check by applying the brake so that the pads engage with the disc. They will lift out of the caliper unit when the front wheel is removed. See Section 5 of this Chapter.

4 If brake action becomes spongy, or if any part of the hydraulic system is dismantled (such as when the hose is replaced, for example) it is necessary to bleed the system in order to remove all traces of air. The following procedure should be followed:

1   Attach a tube to the bleed valve at the top of the caliper unit, after removing the dust cap. It is preferable to use a transparent plastics tube, so that the presence of air bubbles is seen more readily.

2   The far end of the tube should rest in a small bottle so that it is submerged in hydraulic fluid. This is essential, to prevent air from passing back into the system. In consequence, the end of the tube must remain submerged at all times.

3   Check that the reservoir on the handlebars is full of fluid and replace the cap to keep the fluid clean.

4   If spongy brake action necessitates the bleeding operation, squeeze and release the brake lever several times in rapid succession, to allow the pressure in the system to build up. Then open the bleed valve by unscrewing it one complete turn whilst maintaining pressure on the lever. This is a two-person operation. Squeeze the lever fully until it meets the handlebar, then close the bleed valve. If parts of the system have been replaced, the bleed valve can be opened from the beginning and the brake lever worked until fluid issues from the bleed tube. Note that it may be necessary to top up the reservoir during this operation; if it empties, air will enter the system and the whole operation will have to be repeated.

5   Repeat operation 4 until bubbles disappear from the bleed tube. Close the bleed valve fully, remove the bleed tube and replace the dust cap.

6   Check the level in the reservoir and top up if necessary. Never use the fluid that has drained into the bottle at the end of the bleed tube because this contains air bubbles and will re-introduce air into the system. It must stand for 24 hours before it can be re-used.

7   Refit the diaphragm and diaphragm plate and tighten the reservoir cap securely.

8   Do not spill hydraulic fluid on the cycle parts. It is a very effective paint stripper!

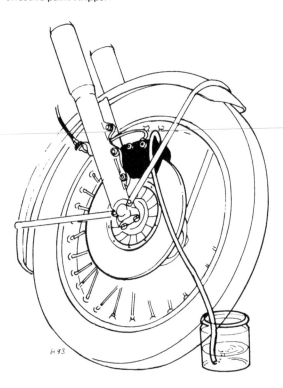

Fig.5 2. Bleeding front disc brake

5   Replacing the pads and overhauling the caliper unit

Model GT 250K only.

1   Remove the front wheel by following the procedure described in Chapter 4, Section 2.6. Rotate the friction pads slightly and withdraw them from the caliper unit.

2   Inspect the friction pads closely and replace them both if the limit level of wear is approached, as described in paragraph 3 of the preceding Section. If there is any doubt whatsoever about their condition, they should be replaced as a pair.

3   Clean the recesses into which the pads fit and the exposed ends of the pistons that actuate them. Use only a small, soft brush and NOT solvent or a wire brush. Smear the piston faces and the brake pad recesses with hydraulic fluid, to act as a lubricant. Only sparing lubrication is required.

4   Remove the reservoir cap, diaphragm plate and diaphragm to check whether the level of fluid rises as the pistons are pushed back into the recesses. It may be necessary to syphon some fluid out of the reservoir prior to this operation, to prevent overflowing. If the pistons do not move freely, the caliper must be removed from the machine and overhauled. Because damage of some kind is inevitable the cause of piston seizures, it is best to entrust the repair of replacement of the unit to a Suzuki repair specialist.

6   Removing and replacing the disc

Model GT 250K only.

1   It is unlikely that the disc will require attention unless it becomes badly scored and braking efficiency is reduced.

2   To remove the disc, first detach the front wheel complete from the forks, as described in Chapter 4, Section 2.6. The disc is bolted to the right-hand side of the wheel hub by six bolts, each pair having a common tab washer. Bend back the tab washer and remove the bolts, to release the disc.

3   Replace the disc by reversing the dismantling procedure. Make sure all the bolts are tightened fully and the tab washers are bent back into position.

7   Master cylinder - examination and renovation

1   The master cylinder is unlikely to give trouble unless the machine has been stored for a lengthy period or until a considerable mileage has been covered. The usual signs of trouble are leakage of hydraulic fluid and a gradual fall in the fluid reservoir content.

2   To gain full access to the master cylinder, commence the dismantling operation by attaching a bleed tube to the caliper unit bleed nipple. Open the bleed nipple one complete turn, then operate the front brake lever until all fluid is pumped out of the reservoir. Close the bleed nipple, detach the tube and store the fluid in a closed container, for subsequent re-use.

3   Detach the hose and also the stop lamp switch (if fitted). Remove the handlebar lever pivot bolt and the lever itself.

4   Access is now available to the piston and the cylinder and it is possible to remove the piston assembly, together with all the relevant seals. Take note of the way in which the seals are arranged because they must be replaced in the same order. Failure to observe this necessity will result in brake failure.

5   Clean the master cylinder and piston with either hydraulic fluid or alcohol. On no account use either abrasives or other solvents such as petrol. If any signs of wear or damage are evident, replacement is necessary. It is not practicable to reclaim either the piston or the cylinder bore.

6   Soak the new replacement seals in hydraulic fluid for about 15 minutes prior to replacement, then reassemble the parts IN EXACTLY THE SAME ORDER, using the reversal of the dismantling procedure. Lubricate with hydraulic fluid and make sure the feather edges of the various seals are not damaged.

7 Refit the assembled master cylinder unit to the handlebar, and reconnect the handlebar lever, hose, stop lamp etc. Refill the reservoir with hydraulic fluid and bleed the entire system by following the procedure detailed in Section 4.4 of this Chapter.
8 Check that the brake is working correctly before taking the machine on the road, to restore pressure and align the pads correctly. Use the brake gently for the first 50 miles or so to enable all the new parts to bed down correctly.

8 Wheel bearings - examination and replacement

1 Access is available to the wheel bearings when the brake plate has been removed. The left-hand wheel bearing is exposed when the brake plate is lifted away; it is protected by an oil seal when the brake plate is reassembled, which remains captive with the brake plate.
2 Lay the wheel on the ground, with the brake drum uppermost and with a drift of the correct diameter, drive out the distance piece that fits within the centre of the left-hand wheel bearing. If this distance piece is driven to the right it will bring with it the right-hand dust cover, spacer, oil seal and bearing. It will be necessary to support the wheel around the outer perimeter for these parts to be driven clear of the hub.
3 Invert the wheel and drive out the left-hand bearing, using a larger diameter drift.
4 Remove all the old grease from the hub and bearings, giving the latter a final wash in petrol. Check the bearings for play or signs of roughness when they are turned. If there is any doubt about their condition, replace them.
5 Before replacing the bearings, first pack the hub with new grease. Then drive them back into position with the same drifts, not forgetting the distance piece that is located between the two bearing centres. Fit the replacement oil seal in front of the bearing on the right-hand side, also the spacer and dust cover.

9 Front wheel - reassembly and replacement

All models except GT 250K

1 Place the front brake plate and brake assembly in the brake drum and align the wheel so that the slot in the brake plate engages with the projection on the lower left-hand fork leg. This acts as the anchorage for the front brake plate.
2 Align the wheel so that the front wheel spindle can be inserted from the right. It may be necessary to spring apart the bottom fork legs a small amount, so that the boss of the hub will locate with the lower right-hand fork leg. Push the spindle home and screw it into the left-hand fork leg using a spanner across the end flats, until the spindle is fully tightened. Lock the spindle in position by re-tightening the clamp bolt through the lower right-hand fork leg.
3 Spin the wheel to check that it moves freely, then attach the front brake cable and the speedometer drive cable. Check that the brake functions correctly, particularly if the brake operating arms have been removed and replaced. If necessary, readjust the brake by following the procedure described in Section 12 of this Chapter.

10 Rear wheel - examination, removal and renovation

1 Place the machine on the centre stand, so that the rear wheel is raised clear of the ground. Check for rim alignment, damage to the rim and loose or broken spokes by following the procedure relating to the front wheel, in the preceding Section.

2 To remove the rear wheel, use the procedure recommended in Section 10.3 of Chapter 4, or, if there is no necessity to disturb either the rear sprocket or the final drive chain, an amended procedure as follows:
Detach the rear brake torque arm from the brake plate by removing the spring clip and withdrawing the securing nut, washer and bolt. Remove also the rear brake cable by unscrewing the adjuster and freeing the cable from the rear brake plate. Then withdraw the wheel from the sprocket and chain assembly by unscrewing the inner of the two left-hand wheel nuts and removing the spindle. When the right-hand distance piece is removed, there will be sufficient clearance for the wheel to be disengaged from the shock absorber vanes of the sprocket, and pulled clear from the frame.
3 The rear brake plate and brake assembly can be withdrawn from the right-hand side of the wheel hub.
4 The rear wheel bearings are also a drive fit in the hub, separated by a spacer. Use a similar technique for removing, greasing and replacing the bearings to that adopted for the front wheel (Section 8 of this Chapter).

8.2. Distance piece separates the wheel bearings

8.3. Left-hand bearing drives out last

10.2. Chain need not be removed if wheel only has to be detached

10.2a Removal of adjuster frees rear brake connection

10.4. Rear wheel bearings are also a drive fit in hub

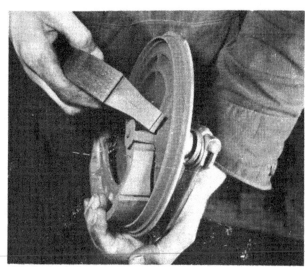

1.2. Brake shoes are separated by pulling off as shown

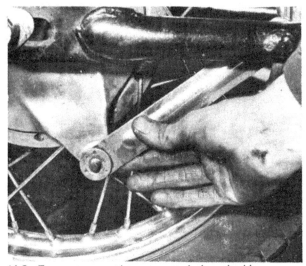

11.2a. Torque arm must be reconnected when wheel is replaced

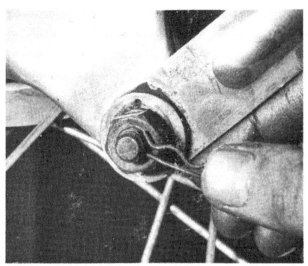

11.2b Do not omit spring clip. If arm works loose, accident will result

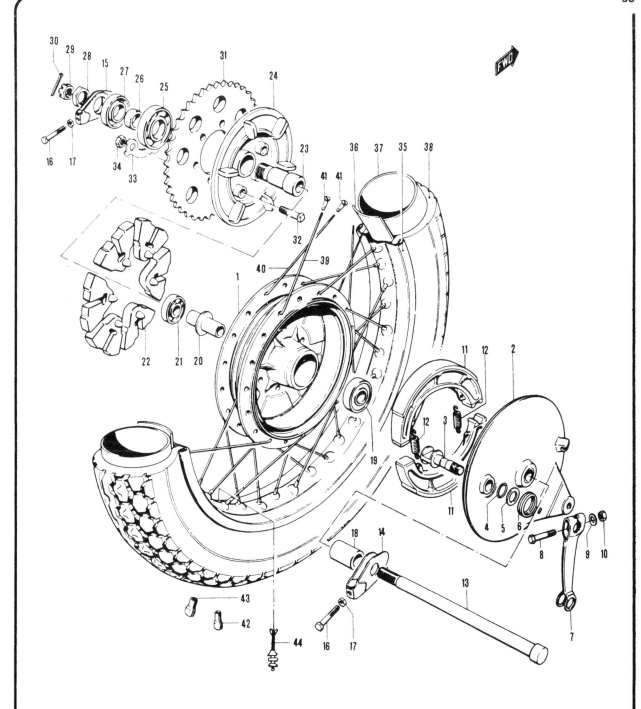

FIG.5.3. REAR WHEEL

1	Rear wheel hub and brake drum	12	Brake shoe spring - 2 off
2	Rear brake plate	13	Rear wheel spindle
3	Brake operating cam	14	Right hand chain adjuster
4	'O' ring seal for brake cam	15	Left hand chain adjuster
5	Brake cam washer	16	Chain adjsuter drawbolt - 2 off
6	Brake cam return spring	17	Drawbolt locknut - 2 off
7	Brake operating lever	18	Rear wheel spindle spacer
8	Brake lever pinch bolt	19	Right hand wheel bearing
9	Flat washer	20	Hub bearing spacer
10	Nut	21	Left hand wheel bearing
11	Brake shoe - 2 off	22	Cush drive rubbers
		23	Sprocket bearing shaft

24	Sprocket cush drive plate
25	Sprocket shaft bearing
26	Sprocket plate spacer
27	Sprocket bearing oil seal
28	Sprocket bearing shaft nut
29	Wheel spindle nut (castellated)
30	Split pin
31	Rear sprocket
32	Rear sprocket retaining bolts - 6 off
33	Sprocket tab washers - 3 off

34	Sprocket retaining nuts - 6 off
35	Wheel rim
36	Rim tape
37	Inner tube
38	Rear tyre
39	Spoke - outer - 18 off
40	Spoke - inner - 18 off
41	Spoke nipple - 36 off
42	Balance weight - short
43	Balance weight - long
44	Inner tube valve insert

11 Rear brake assembly - examination, renovation and re-assembly

1 The rear brake assembly complete with brake plate can be withdrawn from the rear wheel after the wheel spindle has been pulled out, the distance piece removed and the wheel pulled clear of the rear forks. The preceding Section describes a simplified method of wheel removal if the wheel alone is to be removed.

2 If it is necessary to dismantle the rear brake assembly, follow the procedure described in Section 3 of this Chapter that applies to the front wheel. Note that the rear brake is of the single leading shoe type and therefore differs slightly in construction.

12 Adjusting the twin leading shoe front brake

1 If the front brake adjustment is correct, there should be a clearance of not less than 0.8 to 1.2 inches (20 - 30 mm) between the brake lever and the twist grip when the brake is applied fully.

2 Adjustment is effected by turning the adjuster nut in the end of the handlebar lever inwards to increase the clearance and outwards to decrease the clearance, or vice-versa, if the adjuster on the longer of the two brake operating arms is used.

3 The screwed operating rod that joins the two brake operating arms of the front brake should not require attention unless the setting has been disturbed. It is imperative that the leading edge of each brake shoe comes into contact with the brake drum simultaneously, if maximum braking efficiency is to be achieved. Check by detaching clevis pin from the eye of one end of the threaded rod, so that the brake operating arms can be applied independently. Operate each arm at a time and note when the brake shoe first commences to touch the brake drum, with the wheel spinning. Make a mark to show the exact position of each operating arm when this initial contact is made. Replace the clevis pin and check that the marks correspond when the brake is applied in similar fashion. If they do not, withdraw the clevis pin and use the adjuster to either increase or decrease the length of the rod until the marks correspond exactly. Replace the clevis pin and do not omit the split pin through the end that retains it in position. Re-check the brake lever adjustment before the machine is tried on the road.

4 Check that the brake pulls off correctly when the handlebar lever is released. Sluggish action is usually due to a poorly lubricated brake cable, broken return springs or a tendency for the brake operating cams to bind in their bushes. Dragging brakes affect engine performance and can cause severe over-heating of both the brake shoes and wheel bearings.

13 Adjusting the rear brake

1 If the adjustment of the rear brake is correct, the brake pedal will have a travel of from 0.8 to 1.2 inches (20 - 30 mm) Before the amount of travel is adjusted, the brake pedal position should be set so that the pedal is in the best position for quick operation.

2 The height of the brake pedal is determined by the adjuster at the end of the brake cable, where it joins the pedal arm. If the adjuster is screwed inwards, the pedal height is raised and vice-versa.

3 The length of travel is controlled by the adjuster at the end of the brake operating arm. If the nut is screwed inwards, travel is decreased and vice-versa.

4 Note that it may be necessary to readjust the height of the stop lamp switch if the pedal height has been altered to any marked extent. Refer to Chapter 6, for further details.

12 2. Screwed adjuster facilitates adjustment of twin leading shoe brake

14 Cush drive assembly - examination and renovation

1 The cush drive assembly is contained within the left-hand side of the rear wheel hub. It comprises a set of synthetic rubber buffers housed within a series of vanes cast in the hub shell. A plate attached to the rear wheel sprocket has four cast-in dogs that engage with slots in these rubbers, when the wheel is replaced in the frame. The drive to the rear wheel is transmitted via these rubbers, which cushion any surges or roughness in the drive which would otherwise convey the impression of harshness.

2 Examine the rubbers for signs of damage or general deterioration. Replace the rubbers if there is any doubt about their condition; they are held in place by moulded-in pegs on the back that press through holes in the wheel hub shell.

15 Rear wheel sprocket - removal, examination and replacement

1 The rear wheel sprocket assembly can be removed as a separate unit after the rear wheel has been detached from the frame as described in Section 10.2 of this Chapter. Alternatively it can be removed attached to the rear wheel if the procedure described in Section 10.3 of Chapter 4 is followed. In the latter case, the sprocket complete with the sprocket drum shaft and bearing will pull away from the cush drive when the wheel has been detached from the frame.

2 Check the condition of the sprocket teeth. If they are hooked, chipped or badly worn, the sprocket should be replaced. It is retained to the cush drive plate by six nuts and tab washers, which must be removed.

3 It is considered bad practice to replace one sprocket on its own. The final drive sprockets should always be renewed as a pair and a new chain fitted, otherwise rapid wear will necessitate even earlier replacement on the next occasion.

4 An additional bearing is located within the cush drive plate, which supports the sprocket drum shaft into which the rear wheel spindle fits. In common with the wheel bearings, this bearing is of the journal ball type and when wear occurs, the rear wheel sprocket will give the appearance of being slack on its mounting bolts. The bearing is a tight push fit on the sprocket drum shaft and is preceded by an oil seal that excludes road grit and water.

5 Remove the oil seal and bearing and wash out the latter to eliminate all traces of the old grease. If the bearing has any play or runs roughly, it must be replaced.

6 Prior to reassembly, the bearing should be repacked with grease and pushed back onto the shaft, followed by the oil seal. Replace the rear wheel assembly by reversing either of the methods adopted for its removal, whichever is appropriate.

14.1. Cush drive rubbers are contained within rear wheel hub

15.1. Commence by removing chain, spring link positioned at rear wheel sprocket

15.2. Good example of a badly worn sprocket and chain

15.4. Bearing in sprocket has spacer for sprocket drum shaft, then.......

15.4a.......oil seal which protects......

15.4b.......bearing from road grit and water

15.6. Repack bearing with grease before replacement

16.1. Drawbolt adjusters provide means of chain adjustment

15.6a. Sprocket drum shaft pushes into position from inside of cush drive housing

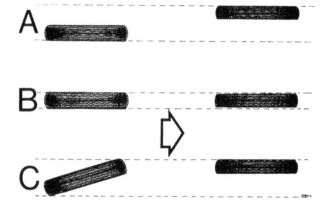

Fig. 5.4. Checking wheel alignment

A & C – Incorrect   B – Correct

15.6b Do not omit split pin from end of spindle when completing reassembly

16 Final drive chain - examination and lubrication

1   The final drive chain is fully exposed, with only a light chain-guard over the top run. Periodically the tension will need to be readjusted, to compensate for wear. This is accomplished by slackening the rear wheel nuts after the machine has been placed on the centre stand and drawing the wheel backwards by means of the drawbolt adjusters in the fork ends. The torque arm bolt on the rear brake plate must also be slackened during this operation.

2   The chain is in correct tension if there is from 0.6 to 0.8 inches (15 - 20 mm) of slack in the middle of the lower run. Always check when the chain is at its tightest point; a chain rarely wears evenly during service.

3   Always adjust the drawbolts an equal amount in order to preserve wheel alignment. The fork ends are marked with a series of horizontal lines above the adjusters, to provide a visual check. If desired, wheel alignment can be checked by running a plank of wood parallel to the machine, so that it touches both walls of the rear tyre. If wheel alignment is correct, it should be equidistant from either side of the front wheel tyre, when tested on both sides of the rear wheel. It will not touch the front wheel tyre because this tyre is of smaller cross section. See accompanying diagram.

4 Do not run the chain overtight to compensate for uneven wear. A tight chain will place excessive stresses on the gearbox and rear wheel bearings, leading to their early failure. It will also absorb a surprising amount of power.

5 After a period of running, the chain will require lubrication. Lack of oil will accelerate wear of both chain and sprockets and lead to harsh transmission. The application of engine oil will act as a temporary expedient, but it is preferable to remove the chain and immerse it in a molten lubricant such as "Linklyfe" or "Chainguard", after it has been cleaned in a paraffin bath. These latter lubricants achieve better penetration of the chain links and rollers and are less likely to be thrown off when the chain is in motion.

6 To check whether the chain requires replacement, lay it lengthwise in a straight line and compress it endwise until all the play is taken up. Anchor one end and pull on the other in order to take up the end play in the opposite direction. If the chain extends by more than the distance between two adjacent rollers, it should be replaced in conjunction with the sprockets. Note that this check should be made AFTER the chain has been washed out, but before any lubricant is applied, otherwise the lubricant will take up some of the play.

7 When replacing the chain, make sure the spring link is seated correctly, with the closed end facing the direction of travel.

17 Tyres - removal and replacement

1 At some time or other the need will arise to remove and replace the tyres, either as the result of a puncture or because a replacement is required to offset wear. To the inexperienced, tyre changing represents a formidable task yet if a few simple rules are observed and the technique learned, the whole operation is surprisingly simple.

2 To remove the tyre from either wheel, first detach the wheel from the machine by following the procedure in Chapter 4, Section 2.6 or Section 6.2 of this Chapter, depending on whether the front or the rear wheel is involved. Deflate the tyre by removing the valve insert and when it is fully deflated, push the bead of the tyre away from the wheel rim on both sides so that the bead enters the centre well of the rim. Remove the locking cap and push the tyre valve into the tyre itself.

3 Insert a tyre lever close to the valve and lever the edge of the tyre over the outside of the wheel rim. Very little force should be necessary; if resistance is encountered it is probably due to the fact that the tyre beads have not entered the well of the wheel rim all the way round the tyre.

4 Once the tyre has been edged over the wheel rim, it is easy to work around the wheel rim so that the tyre is completely free on one side. At this stage, the inner tube can be removed.

5 Working from the other side of the wheel, ease the other edge of the tyre over the outside of the wheel rim that is furthest away. Continue to work around the rim until the tyre is free completely from the rim.

6 If a puncture has necessitated the removal of the tyre, re-inflate the inner tube and immerse it in a bowl of water to trace the source of the leak. Mark its position and deflate the tube. Dry the tube and clean the area around the puncture with a petrol-soaked rag. When the surface has dried, apply the rubber solution and allow this to dry before removing the backing from the patch and applying the patch to the surface.

7 It is best to use a patch of the self-vulcanising type, which will form a very permanent repair. Note that it may be necessary to remove a protective covering from the top surface of the patch, after it has sealed in position. Inner tubes made from synthetic rubber may require a special type of patch and adhesive, if a satisfactory bond is to be achieved.

8 Before replacing the tyre, check the inside to make sure the agent that caused the puncture is not trapped. Check also the outside of the tyre, particularly the tread area, to make sure nothing is trapped that may cause a further puncture.

9 If the inner tube has been patched on a number of past occasions, or if there is a tear or large hole, it is preferable to discard it and fit a replacement. Sudden deflation may cause an accident, particularly if it occurs with the front wheel.

10 To replace the tyre, inflate the inner tube sufficiently for it to assume a circular shape but only just. Then push it into the tyre so that it is enclosed completely. Lay the tyre on the wheel at an angle and insert the valve through the rim tape and the hole in the wheel rim. Attach the locking cap on the first few threads, sufficient to hold the valve captive in its correct location.

11 Starting at the point furthest from the valve, push the tyre bead over the edge of the wheel rim until it is located in the central well. Continue to work around the tyre in this fashion until the whole of one side of the tyre is on the rim. It may be necessary to use a tyre lever during the final stages.

12 Make sure there is no pull on the tyre valve and again commencing with the area furthest from the valve, ease the other bead of the tyre over the edge of the rim. Finish with the area close to the valve, pushing the valve up into the tyre until the locking cap touches the rim. This will ensure the inner tube is not trapped when the last section of the bead is edged over the rim with a tyre lever.

13 Check that the inner tube is not trapped at any point. Reinflate the inner tube, and check that the tyre is seating correctly around the wheel rim. There should be a thin rib moulded around the wall of the tyre on both sides, which should be equidistant from the wheel rim at all points. If the tyre is unevenly located on the rim, try bouncing the wheel when the tyre is at the recommended pressure. It is probable that one of the beads has not pulled clear of the centre well.

14 Always run the tyres at the recommended pressures and never under or over-inflate. The correct pressures for solo use are 22 psi front and 26 psi rear. If a pillion passenger is carried, increase the rear tyre pressure only to 30 psi.

15 Tyre replacement is aided by dusting the side walls, particularly in the vicinity of the beads, with a liberal coating of french chalk. Washing-up liquid can also be used to good effect, but this has the disadvantage of causing the inner surfaces of the wheel rim to rust.

16 Never replace the inner tube and tyre without the rim tape in position. If this precaution is overlooked there is good chance of the ends of the spoke nipples chafing the inner tube and causing a crop of punctures.

17 Never fit a tyre that has a damaged tread or side walls. Apart from the legal aspects, there is a very great risk of a blow-out, which can have serious consequences on any two-wheel vehicle.

18 Tyre valves rarely give trouble, but it is always advisable to check whether the valve itself is leaking before removing the tyre. Do not forget to fit the dust cap, which forms an effective second seal.

Fault diagnosis - Wheels, brakes and tyres

Symptom	Reason/s	Remedy
Handlebars oscillate at low speeds	Buckled front wheel Incorrectly fitted front tyre	Remove wheel for specialist attention. Check whether line around bead is equidistant from rim.
Forks 'hammer' at high speeds	Front wheel out of balance	Add weights until wheel will stop in any position.
Brakes grab, locking wheel	Ends of brake shoes not chamfered	Remove brake shoes and chamfer ends.
Brakes feel spongy	Stretched brake operating cables, weak pull-off springs. Air in hydraulic fluid (GT250K)	Replace cables and/or springs, after inspection. Bleed brake (GT250K).
Tyres wear more rapidly in middle of tread	Over inflation	Check pressures and run at recommended settings.
Tyres wear rapidly at outer edges of tread	Under-inflation	Ditto.

Tyre changing sequence - tubed tyres

 A Deflate tyre. After pushing tyre beads away from rim flanges push tyre bead into well of rim at point opposite valve. Insert tyre lever adjacent to valve and work bead over edge of rim.

Use two levers to work bead over edge of rim. Note use of rim protectors **B**

C Remove inner tube from tyre

When first bead is clear, remove tyre as shown **D**

E When fitting, partially inflate inner tube and insert in tyre

Work first bead over rim and feed valve through hole in rim. Partially screw on retaining nut to hold valve in place. **F**

G Check that inner tube is positioned correctly and work second bead over rim using tyre levers. Start at a point opposite valve.

Work final area of bead over rim whilst pushing valve inwards to ensure that inner tube is not trapped **H**

Chapter 6 Electrical system

Contents

Specifications

Battery

Type	Lead acid
Make	Yuasa 12N5 - 3
Voltage	12 volts
Capacity	5 amp. hr.

Alternator

Output	1.5 - 2.5 amps at 8000 rpm (daytime running - light electrical load)
	2 - 3 amps at 8000 rpm (night running - heavy electrical load)

Bulbs

Main headlamp	35/25 W Pre-focus
Parking lamp	3W bayonet fitting
Tail/stop lamp	7/23W offset pins
Speedometer lamp	3W bayonet fitting
Tachometer lamp	3W bayonet fitting
Neutral indicator lamp	3W bayonet fitting
Headlamp beam indicator lamp	3W bayonet fitting
Flashing indicators lamp	1.5W bayonet fitting
Flashing indicator lamps	23W each bayonet fitting
	All bulbs 12 volt rating

1 General description

1 The 250 cc and 350 cc Suzuki twins are fitted with a 12 volt electrical system. The circuit comprises a crankshaft-driven rotating magnet alternator which has a stator with six coils, each pair coupled in series. During daytime running, only one set of coils is used because the only electrical demand is from the ignition circuit and the occasional use of the stop lamp. At night, all three sets of coils are used in order to meet the additional load of the lighting equipment. The coils work in parallel, to supply the extra current.

2 The output from the alternator is a.c. hence a rectifier is included in the circuit to covert this current to d.c. for charging the 12 volt, five amp hr battery. The daytime charging rate is within the 1.5 - 2.5 ampere range; at night the rate increases to within the 2 - 3 ampere range. These are the peak readings at 8,000 rpm engine speed.

2 Crankshaft alternator - checking the output

1 As explained in Chapter 3, Section 2 the output from the alternator, can be checked only with specialised test equipment of the multi-meter type. If the performance of the alternator is in any way suspect, it should be checked by either a Suzuki agent or an auto-electrical specialist.

3 Battery - inspection and maintenance

1 A Yuasa type 12N5 - 3 battery is fitted as standard. This battery is of the lead-acid type and has a capacity of 5 amp hrs.

2 The transparent case of the battery allows the upper and lower levels of the electrolyte to be observed without need to remove the battery. Maintenance is normally limited to keeping the electrolyte level between the prescribed upper and lower limits and making sure the vent tube is not blocked. The lead

plates and their separators can be seen through the transparent case, a further guide to the condition of the battery.

3 Unless acid is spilt, as may occur if the machine falls over, the electrolyte should always be topped up with distilled water, to restore the correct level. If acid is spilt on any part of the machine, it should be neutralised with an alkali such as washing soda and washed away with plenty of water, otherwise serious corrosion will occur. Top up with sulphuric acid of the correct specific gravity (1.260 - 1.280) only when spillage has occurred. Check that the vent pipe is well clear of the frame tubes or any of the other cycle parts.

4 It is seldom practicable to repair a cracked case because the acid in the joint prevents the formation of an effective seal. It is always best to replace a cracked battery, especially in view of the corrosion that will be caused by acid leakage.

5 If the machine is laid up for a period, it is advisable to remove the battery and give it a 'refresher' charge every six weeks or so from a battery charger. If the battery is permitted to discharge completely, the plates will sulphate and render the battery useless.

4 Battery - charging procedure

1 The normal charging rate for the 5 amp hr battery fitted to the Suzuki twins is 0.2 amps. A more rapid charge can be given in an emergency, in which case the charging rate can be raised to 0.6 - 1 amp. The higher charge rate should be avoided if possible, because this will eventually shorten the working life of the battery.

2 Make sure the charger connections to the battery are correct; red to positive and black to negative. It is preferable to remove the battery from the machine during the charging operation and to remove the vent plug from each cell.

5 Silicon rectifier - general description

1 The function of the silicon rectifier is to convert the a.c. current produced by the alternator to d.c. so that it can be used to charge the battery. The rectifier is of the full wave type.

2 The rectifier is located in a position where it is not directly exposed to water or battery acid, which may cause it to malfunction. The question of access is of little importance because the rectifier is unlikely to give trouble during normal service. It is not practicable to repair a damaged rectifier. If the unit malfunctions, it must be replaced.

3 Damage to the rectifier will occur if the machine is run without the battery for any period of time. A high voltage will develop in the absence of any load on the electrical coils, which will cause a reverse flow of current and consequent damage to the rectifier cells. Reverse connection of the battery will also have the same effect.

4 It is not possible to check whether the rectifier is functioning correctly without the appropriate test equipment. A Suzuki agent or an auto-electrical specialist are best qualified to advise.

5 Do not loosen the rectifier locking nut, or in any way damage the surfaces of the assembly. As such action may cause the coating over the electrodes to peel and destroy the working action.

6 Fuse - location and replacement

1 A fuse is incorporated in the electrical system to give protection from a sudden overload, as may occur during a short circuit. It is found within a fuse holder that forms part of the wiring snap connections, close to the battery. A transparent plastics bag attached to the wiring carries a spare fuse, for use in an emergency. The fuse is rated at 15 amps.

2 If a fuse blows, it should be replaced, after checking to ensure that no obvious short circuit has occurred. If the second fuse blows shortly afterwards, the electrical circuit must be checked thoroughly, to trace the fault.

3 When a fuse blows whilst the machine is running and no spare is available, a 'get you home' remedy is to remove the blown fuse and wrap it in silver paper before replacing it in the fuseholder. The silver paper will restore electrical continuity by bridging the broken fuse wire. This expedient should never be used if there is evidence of a short circuit or other major electrical fault, otherwise more serious damage will be caused. Replace the blown fuse at the earliest possible opportunity, to restore full circuit protection.

7 Headlamp - replacing bulbs and adjusting beam height

1 To remove the headlamp rim, detach the two small cross head screws in the lower front portion of the headlamp shell (one near each fork leg). The headlamp rim can now be pulled away from the shell and lifted off when it has cleared the lip at the top.

2 The main bulb is of the double-filament type, to give a dipped beam facility. The bulb holder is attached to the reflector by a rubber sleeve, which fits around the flange in the reflector and the flange of the bulb holder itself. The bulb is rated at 12 volts, 35/25W.

3 It is not necessary to refocus the headlamp when a new bulb is fitted because the bulbs used are of the pre-focus type. To release the main headlamp bulb, press and twist it in the holder.

4 The pilot lamp bulb holder is a bayonet fitting in the reflector, below the main bulb. The bulb holder is protected by a rubber sleeve. Used for parking purposes, only, the bulb has a 12 volt, 3W rating.

5 Beam alignment is adjusted by means of a small screw through the left-hand side of the headlamp rim, just below the telescopic fork lug to which the headlamp shell is attached. The screw passes through a plate attached to the back of the reflector, into a threaded nylon insert. By turning the screw, the headlamp beam can be ranged to either the right or the left, in the horizontal plane.

6 Beam height is adjusted by slackening the two bolts that retain the headlamp shell in position (through the lugs from the telescopic forks) and tilting the shell either upwards or downwards before retightening.

7 To check the headlamp alignment, place the machine on level ground facing a wall 25 feet distant, with the rider seated normally. The height of the beam centre should be equal to that of the height of the centre of the headlamp from the ground, when the dip switch is in the 'full on' position. The concentrated area of light should be centrally disposed. Adjustments in either direction are made as detailed in the preceding paragraphs. Note that a different setting for the beam height will be required when a pillion passenger is carried.

8 The above instructions for beam setting relate to the requirements of the United Kingdom's Transport Lighting Regulations. Other settings may be required in countries other than the UK.

7.1. Remove headlamp rim by detaching cross head screws from bottom of shell

7.2. Main bulb is attached to reflector unit by rubber sleeve

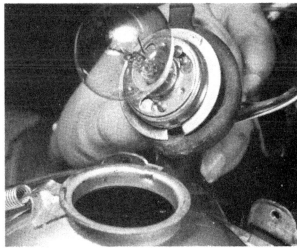

7.3. Twist bulb in holder to release. Bulb is of prefocus type

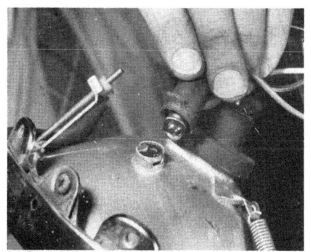

7.4. Pilot bulb holder has bayonet fixing in reflector unit

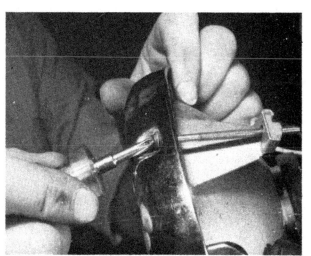

7.5. Beam alignment is effected by adjusting screw in headlamp rim

8 Handlebar dipswitch - examination

1   The dipswitch forms part of the left-hand 'dummy' twist grip and should not normally give trouble. In the event of failure, the switch assembly complete must be replaced; it is not practicable to effect a permanent repair.

9 Stop and tail lamp - replacing bulbs

1   The tail lamp has a twin filament bulb of 12 volt, 7/23W rating, to illuminate the rear number plate and to give visual warning when the rear brake is applied. To gain access to the bulb, remove the two screws that retain the moulded plastics lens cover to the tail lamp assembly, and remove the cover complete with gasket. The bulb has a bayonet fitting, with staggered pins to prevent the bulb contacts from being reversed.
2   If the tail lamp bulb keeps blowing, suspect either vibration in the rear mudguard assembly, or more probably, a poor earth connection.
3   The stop lamp is operated by a stop lamp switch on the right-hand side of the machine, immediately above the brake pedal. It is connected to the pedal by an extension spring, which acts as the operating link. The body of the switch is threaded, so that a limited range of adjustment is available, to determine when the lamp will operate.

10 Flashing indicators

1   The forward flashing indicator lamps are connected to 'stalks' that thread into the alloy top yoke of the telescopic forks. They are retained by a locknut and can be set to any level desired. The rear-facing indicator lamps are also mounted on stalks and thread through a metal plate attached to each side of the rear carrier.
2   In each case, access to the bulb is gained by removing the moulded plastics lens cover. Each bulb is rated at 12 volts, 23W.

9.1. Tail and stop lamp bulb is available after removing plastic lens cover

10.1. Remove screws from yellow lens to gain access to flashing indicator bulbs

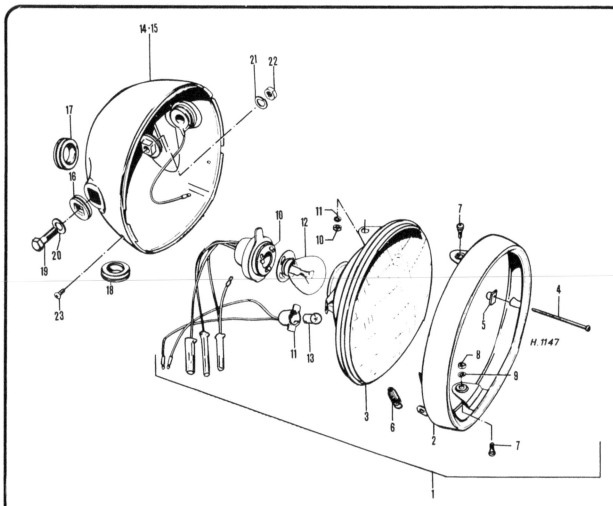

FIG.6.1. HEADLAMP

1 Headlamp assembly	8 Nut - 2 off	bulb	19 Headlamp mounting bolt - 2 off
2 Rim	9 Spring washer - 2 off	14 Headlamp shell - chromium plated	20 Flat washer - 2 off
3 Reflector unit	10 Bulb holder - main bulb	15 Headlamp shell painted	21 Spring washer - 2 off
4 Beam adjusting screw	11 Bulb holder - parking lamp	16 Rubber - 2 off	22 Nut - 2 off
5 Beam adjusting screw nut	12 Headlamp main bulb	17 Grommet	23 Screw - 2 off
6 Bulb holder spring	13 Headlamp parking lamp	18 Grommet	
7 Screw - 2 off			

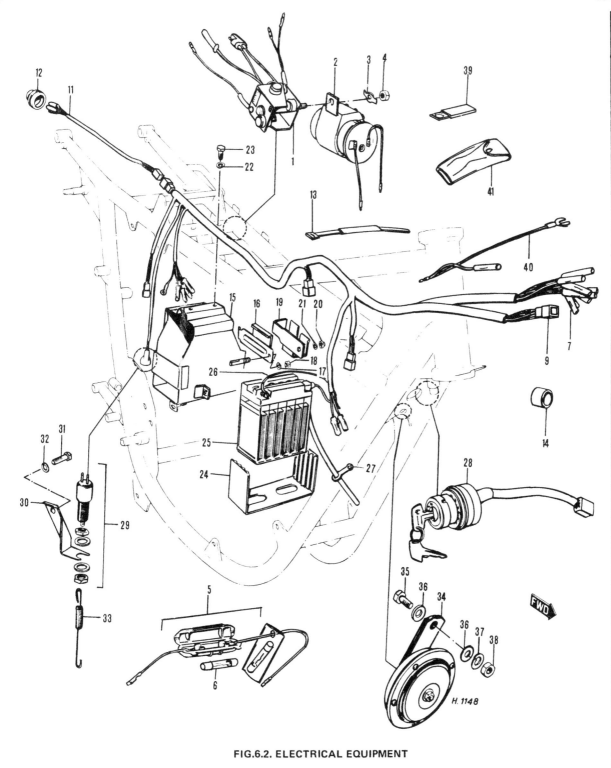

FIG.6.2. ELECTRICAL EQUIPMENT

1	Rectifier assembly	12	Wiring harness grommet	23	Bolt - 2 off	34	Horn
2	Flasher unit	13	Wiring harness clamp	24	Battery cushion pad	35	Bolt
3	Rectifier lock washer	14	Wiring harness ring - 2 off	25	Battery	36	Flat washer - 2 off
4	Nut	15	Battery holder	26	Battery lead wire	37	Spring washer
5	Fuse holder	16	Battery holder clamp	27	Battery vent pipe holder	38	Nut
6	Fuse	17	Spring washer	28	Ignition switch	39	Rectifier wire clamp
7	Wiring harnesses	18	Nut	29	Stop lamp switch	40	Handlebar earth wire
8	Ditto	19	Tool kit retainer	30	Stop lamp switch bracket	41	Tool kit
9	Ditto	20	Nut	31	Bolt		
10	Ditto	21	Spring washer	32	Spring washer		
11	Ditto	22	Spring washer - 2 off	33	Stop lamp switch spring		

11 Flasher unit - location and replacement

1   The flasher unit is bolted to the left-hand side of the rectifier assembly, below and close to the nose of the dual seat.
2   A series of audible clicks will be heard if the flasher unit is functioning correctly. If the unit malfunctions, the usual symptom is one initial flash before the unit goes dead. It will be necessary to replace the flasher unit complete if the fault cannot be attributed to either a burnt-out indicator bulb or a blown fuse. Take great care when handling the unit because it is easily damaged if dropped.

12 Tachometer head - replacement of bulbs

1   The tachometer head houses no less than four bulbs, each of which has an indicating function, apart from the bulb used for internal illumination. The indicating function and the bulb ratings are as follows:

Flashing indicator lamp	12 volt 1.5W
Full headlamp beam indicator lamp	12 volt, 3W
Neutral indicator lamp	12 volt, 3W
Tachometer dial illumination lamp	12 volt, 3W

2   The bulb holders are a push fit into the base of the tachometer head, where they are retained by their outer moulded rubber sleeves. The bulbs have a bayonet fitting.

13 Speedometer head - replacement of bulb

1   The speedometer dial is illuminated by a 12 volt, 3W bulb that is a push fit into the base of the head, as in the case of the matching tachometer.
2   The speedometer illuminating bulb also has a bayonet fitting.

14 Horn - location and examination

1   The horn is suspended from a flexible steel strip mounted on the cross member that joins the two front down tubes of the frame, immediately below the steering head. The flexible steel strip isolates the horn from the undesirable effects of high frequency vibration.
2   The horn has no external means of adjustment. If it malfunctions it must be replaced; it is a statutory requirement that the machine must be fitted with a horn that is in working order.

15 Wiring - layout and examination

1   The wiring harness is colour-coded and will correspond with the accompanying diagram. Where socket connectors are used, they are designed so that reconnection can be made only in the one correct position.
2   Visual inspection will show whether any breaks or frayed outer coverings are giving rise to short circuits. Another source of trouble may be the snap connectors and sockets, where the connector has not been pushed home fully in the outer housing.
3   Intermittent short circuits can often be traced to a chafed wire that passes through or is close to a metal component, such as a frame member. Avoid tight bends in the wire or situations where the wire can become trapped between casings.

16 Ignition and lighting switch

1   The ignition and lighting switch is combined in one unit. It is operated by a key, which cannot be removed when the ignition is switched on.
2   The number stamped on the key will match also the number of the steering head lock and the filler cap lock. A replacement key can be obtained if the number is quoted; if either lock or the ignition switch is changed extra keys will be needed.
3   It is not practicable to repair the ignition switch if it malfunctions. It should be replaced with a new lock and key to suit.

16.1. Ignition switch has four positions, two of which control lights

Fault diagnosis - Electrical system

Symptom	Reason/s	Remedy
Complete electrical failure	Blown fuse	Check wiring and electrical components for short circuit before fitting new 15 amp fuse.
	Isolated battery	Check battery connections, also whether connections show signs of corrosion.
Dim lights, horn inoperative	Discharged battery	Recharge battery with battery charger and check whether alternator is giving correct output (electrical specialist)
Constantly 'blowing' bulbs	Vibration, poor earth connection	Check whether bulb holders are secured correctly. Check earth return or connections to frame.

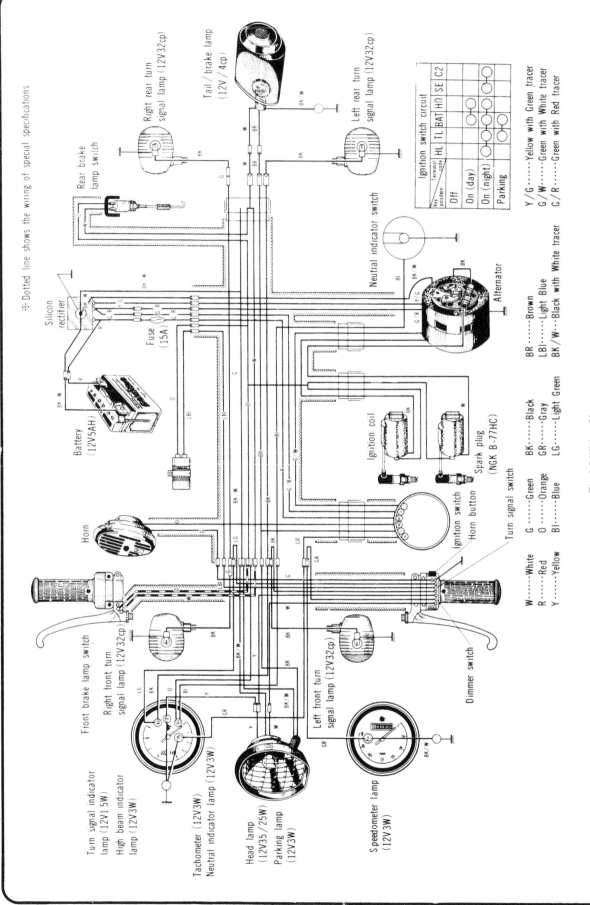

Fig.6.3 Wiring Chart

1977 Suzuki GT250A

Chapter 7 Suzuki GT250M
and GT250A, B and C models

Contents

Specifications

Engine

Piston to cylinder clearance...	0.040–0.050 mm (0.0016–0.0020 in)

Figure based upon piston diameter as measured at 26 mm (1.0) above the skirt (by micrometer reading)

Piston ring end gap	0.15–0.35 mm (0.0059–0.0138 in)
Gearbox oil	1,300 cc (2.8/2.3 US/Imp pts)

Gear ratios

	GT250A	GT250M
First	2.333 : 1 (28/12)	2.333 : 1 (28/12)
Second	1.352 : 1 (23/17)	1.500 : 1 (24/16)
Third	1.050 : 1 (21/20)	1.105 : 1 (21/19)
Fourth	0.905 : 1 (19/21)	0.905 : 1 (19/21)
Fifth	0.782 : 1 (18/23)	0.782 : 1 (18/23)
Top	0.708 : 1 (17/24)	0.708 : 1 (17/24)
Primary reduction	3.050 : 1 (61/20)	3.050 : 1 (61/20)
Final reduction	3.071 : 1 (43/14)	2.929 : 1 (41/14)

Carburettors

	GT250A	GT250M
Type	VM28SS	VM26SH
Main jet:		
right-hand	95	112.5
left-hand	97.5	112.5
Pilot jet	30	25
Jet needle	5CN3–3	5CN3–3
Needle jet	0–2	0–2
Air screw	1½ turns out	1½ turns out
Starter jet	100	80
Cutaway	2.5	2.5
Float level	13.6 mm (0.53 in)	31 mm (1.2 in)
Engine idling speed	1,350 rpm	
Pilot air adjusting screw setting	1–1½ turns back from fully closed	

Ignition system

Sparking plug specification	NGK B–9ES
Sparking plug gap	0.6 – 0.7 mm (0.024 – 0.028 in)
Contact/breaker point	0.35 mm (0.014 in)
Ignition timing:	
M model	$24° \pm 2°$ BTDC (piston travel 2.93 mm 0.115 in)
A, B and C models	$20° \pm 2°$ BTDC (piston travel 2.05 mm 0.081 in)

Frame and forks

Drive chain free play	15–20 mm (0.6–0.8 in)
Fork oil	145cc (4.9/5.1 US/Imp ozs) automatic transmission fluid for each leg

1 General description

1   The crankshaft of the GT250M model is supported by three roller bearings and has a separate middle mainshaft pressed between the two sets of flywheels. The GT250A crankshaft has four roller bearings to support it and has one of its flywheels built integral with the middle mainshaft. This helps to ensure greater rigidity (see Figs 7.1 and 7.2).

2   The new GT250A model no longer has a ram air cover. The new integral cylinder head has greatly improved finning making the ram air cover unnecessary.

3   In the crankcase of the GT250A the oil-ways have been relocated in the upper crankcase, one of them for the left-hand bearing and the other for the centre bearing.

4   The air filter on the GT250A model has been changed from a paper element type to a wet polyurethane foam type, which is claimed to have a much better filtering performance than the older type. It can be removed easily and washed in petrol, which enables it to be reused many times.

5   To accommodate the change in the engine from the two scavenging port design to the four scavenging port design of the GT 250A model, alterations have been made to the carburettor setting. The carburettors are also mounted differently in order to make the fuel level in the float chamber more stable and to ensure the best possible carburation under all conditions. To accomplish this the carburettors are no longer rigidly mounted, a rubber inlet pipe being used to hold the carburettors in a floating manner and thereby reduce engine vibration from being transmitted.

6   The gear ratios for second and third gears differ between the two models (see table in specifications).

7   Two further models, the GT250B and GT250C, were introduced in 1977 and 1978 respectively. These models differ from the GT250A only in minor respects, the greatest changes being in the styling of some cycle parts. Where information is required relating to either of these models, refer to the Specifications or procedure given for the GT250A.

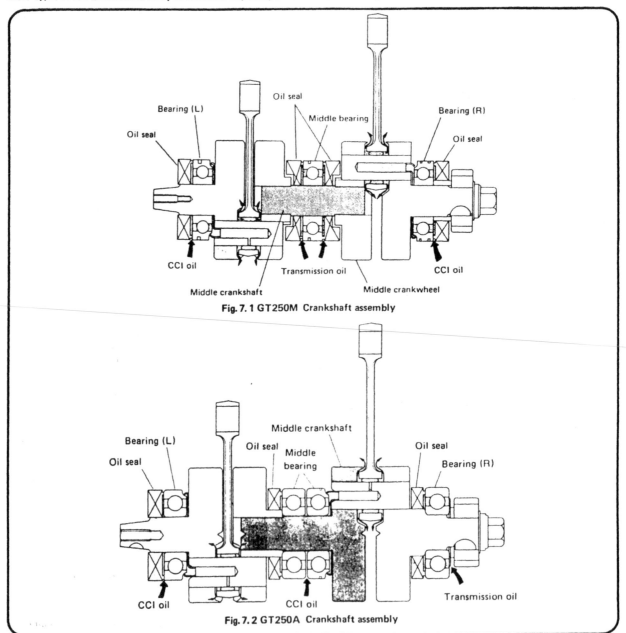

Fig. 7.1 GT250M Crankshaft assembly

Fig. 7.2 GT250A Crankshaft assembly

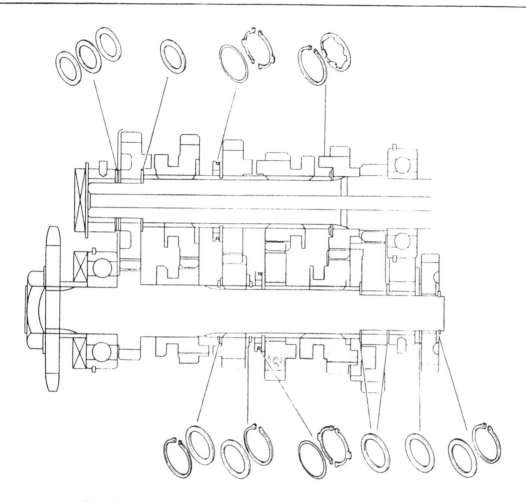

Fig. 7.3 Location of thrust washers and circlips on gear cluster (GT250A model)

2 Adjusting the GT250A Carburettor float level

1 The measurement of the carburettor float level is made between the points shown in fig 7.4, when the carburettor is held upside down. The measurement must be checked when the carburettor is held upside down to ensure that the float arm is free. Bring the float arm into contact with the needle valve then measure the distance, as shown in the diagram, between the mounting face of the float chamber bowl and the float arm. The specified distance is 13.6 mm (0.56in)

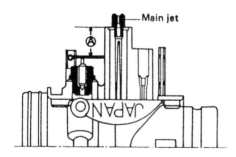

Fig. 7.4 Carburettor float level setting

3 Cleaning the GT250A air filter

1 The polyurethane foam element in the air filter should remain wet with engine oil because a dry foam element would not fulfil its purpose. The element usually requires attention at approximately 2000 mile intervals, or whenever upon inspection it is found to be dirty. The element should be removed and washed thoroughly in clean petrol. Any large particles of dirt or dead flies etc., may be removed by hand. When the element appears to be clean, squeeze out excess petrol (do not twist or wring the element) to make sure it is as dry as possible. Then moisten it with Texaco (or similar) two-stroke oil. The easiest way of ensuring that the oil is evenly distributed throughout the element, is to first soak it in two-stroke oil and then squeeze out the excess. Make sure that no oil drips from the filter before replacing it in the machine.

4 Adjusting the GT250A oil pump

1 The oil pump is adjusted by turning the cable adjuster until the aligning marks coincide, when the throttle valve depression mark is at the upper part of the carburettor alignment port. This latter is achieved by opening the throttle when the ignition is switched off. Be sure to secure the adjuster with the lock nut, after the oil pump setting is correct, if adjustment has proved necessary to make the timing marks coincide.

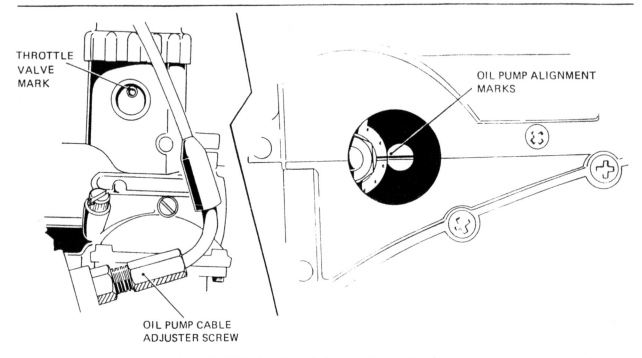

THROTTLE
VALVE
MARK

OIL PUMP ALIGNMENT
MARKS

OIL PUMP CABLE
ADJUSTER SCREW

Fig. 7.5 Carburettor and oil pump adjustment marks

5 Ignition timing

The ignition timing for the later models is as follows:

GT 250M 24° ± 2° BTDC or 2.93 mm (0.115) of piston travel

GT 250A 20° ± 2° BTDC or 2.05 mm (0.081) of piston travel

6 Final drive chain: changing — endless type

1 The GT250A models are fitted with an 'endless' type chain. Despite its title, it should not be assumed that the chain does not need periodic maintenance. On the contrary, in view of the fact that considerable dismantling is necessary to renew the chain, frequent cleaning and lubrication will prolong its life and postpone the inevitable.

2 The chain should be cleaned in situ, and relubricated with one of the proprietary aerosol lubricants designed for this purpose. Engine oil is not really suitable as it tends to be flung off easily. Do not use paraffin as a cleaner and do not over-lubricate. Little and often is advised.

3 This task should be undertaken every 200 miles at least. In wet, dirty conditions this period can be reduced considerably, lubrication being especially important before and after a long run.

4 The amount of chain wear can be assessed after removing the chain from the machine (see following Section). Clean the chain thoroughly in petrol and dry it off with a clean rag (take care not to place the chain anywhere where grit may stick to it, as this may damage both the chain and the sprockets if it is refitted to the bike). Draw a line exactly 1 foot in length on a piece of card. Lay the chain along the length of the line, anchor one end firmly and then compress the chain along the length of the line until all the links are as close to one another as possible. Check that the links are still against the line and mark the position of the end of the chain. Pull the links apart as far as possible, making sure that the chain length is fully extended. Mark the new length of the chain after checking that it is still against the line. The amount it has stretched between the two marks should not be more than ¼ of an inch per foot length. If the chain has stretched more than this, it should be renewed.

7 Endless type chain removal

1 The correct way to remove the chain is as follows:

2 Place the machine securely on its centre stand, and block, if necessary, to raise the rear wheel clear of the ground. The silencers should be detached to give better access, as should the chain guard.

3 Slacken off both chain adjusters, and the wheel spindle nut. Allow the adjusters to drop, exposing the stops at the end of the fork ends. These are each retained by a single bolt and should be removed.

4 Push the wheel forward so that the chain can be disengaged from the sprocket. Allow it to rest to one side. Disconnect the torque arm from the brake plate by removing the single retaining nut. Detach the rear brake cable at the actuating lever.

5 Remove the wheel spindle nut completely, and withdraw the spindle. The wheel can now be drawn clear.

6 Remove the lower mounting bolts from the suspension units, and free the units from their mounting lugs.

7 Remove the pivot shaft nut, and withdraw the shaft. The swinging arm can now be pulled clear and disengaged from the chain.

8 Detach the left-hand footrest and the gearchange pedal, followed by the starter cover and gasket. Remove the engine sprocket cover after first removing its four retaining bolts.

9 Knock back the tabwasher, and remove the engine sprocket nut. The sprocket can be locked in position by bunching the chain against the casing.

10 Pull off the sprocket and disengage the chain. The new chain can be fitted by reversing the removal sequence. Ensure that chain is adjusted correctly. Tighten the wheel spindle.

8 Tracing generator faults.

1 A quick test for tracing generator faults in the GT250 A and M models (which may also be used on earlier models) is as follows. Remove the dualseat and disconnect the green wire with a white tracer from the block connector of the generator harness. For the next step a Suzuki pocket tester or a multimeter is required.

Switch to the 150 Volts AC range and place the two tester terminals across the yellow wire with the green tracer. Start the machine, when the following readings should be obtained:

    30 volts   at 2000 rpm
    60 volts   at 4000 rpm
    90 volts   at 6000 rpm

If the readings obtained are different from these recommendations, suspect one or more of the charging coils of being defective. This will necessitate the substitution of a new stator assembly. The defect should be verified by a Suzuki Service Agent or an auto-electrical specialist before the stator is renewed.

9 Other modifications

1 A more efficient type of silencer is used on the GT250A model.

2 Rear shock absorbers on the GT250M and GT250A models are no different in terms of performance although the GT250A type has no upper cover.

3 The styling of the handlebar grips differs between the M and A models.

4 The pillion footrests have been moved from the swinging arm of the GT250M model to the frame of the GT250A model as an aid to riding comfort.

5 A larger size flashing indicator lens is used on the GT250A model. It is now 80 mm (3.2in) as opposed to 70 mm (2.8in) on the GT250M model.

6 Piston ring sizes are the same but the positioning pegs for the piston rings are in different locations. Those of the GT250A model are 80° apart, and in the case of the GT250M model 40° apart.

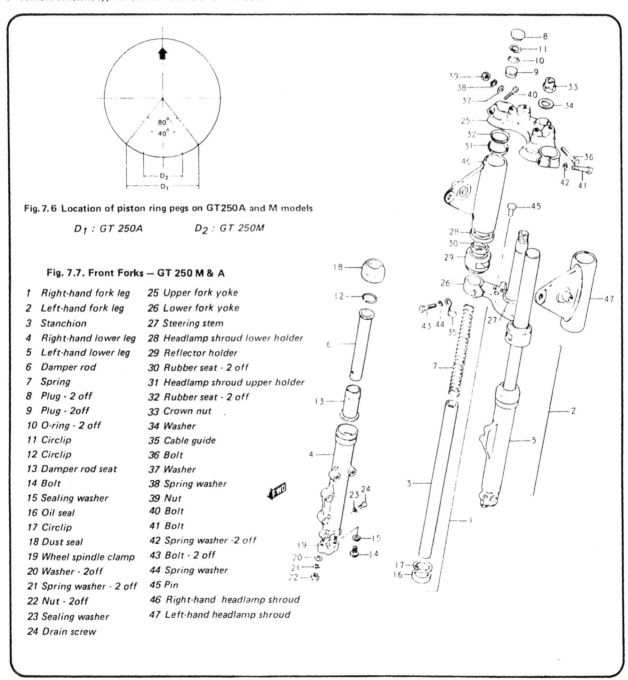

Fig. 7.6 Location of piston ring pegs on GT250A and M models

D_1 : GT 250A D_2 : GT 250M

Fig. 7.7. Front Forks — GT 250 M & A

1 Right-hand fork leg	25 Upper fork yoke
2 Left-hand fork leg	26 Lower fork yoke
3 Stanchion	27 Steering stem
4 Right-hand lower leg	28 Headlamp shroud lower holder
5 Left-hand lower leg	29 Reflector holder
6 Damper rod	30 Rubber seat - 2 off
7 Spring	31 Headlamp shroud upper holder
8 Plug - 2 off	32 Rubber seat - 2 off
9 Plug - 2off	33 Crown nut
10 O-ring - 2 off	34 Washer
11 Circlip	35 Cable guide
12 Circlip	36 Bolt
13 Damper rod seat	37 Washer
14 Bolt	38 Spring washer
15 Sealing washer	39 Nut
16 Oil seal	40 Bolt
17 Circlip	41 Bolt
18 Dust seal	42 Spring washer -2 off
19 Wheel spindle clamp	43 Bolt - 2 off
20 Washer - 2off	44 Spring washer
21 Spring washer - 2 off	45 Pin
22 Nut - 2off	46 Right-hand headlamp shroud
23 Sealing washer	47 Left-hand headlamp shroud
24 Drain screw	

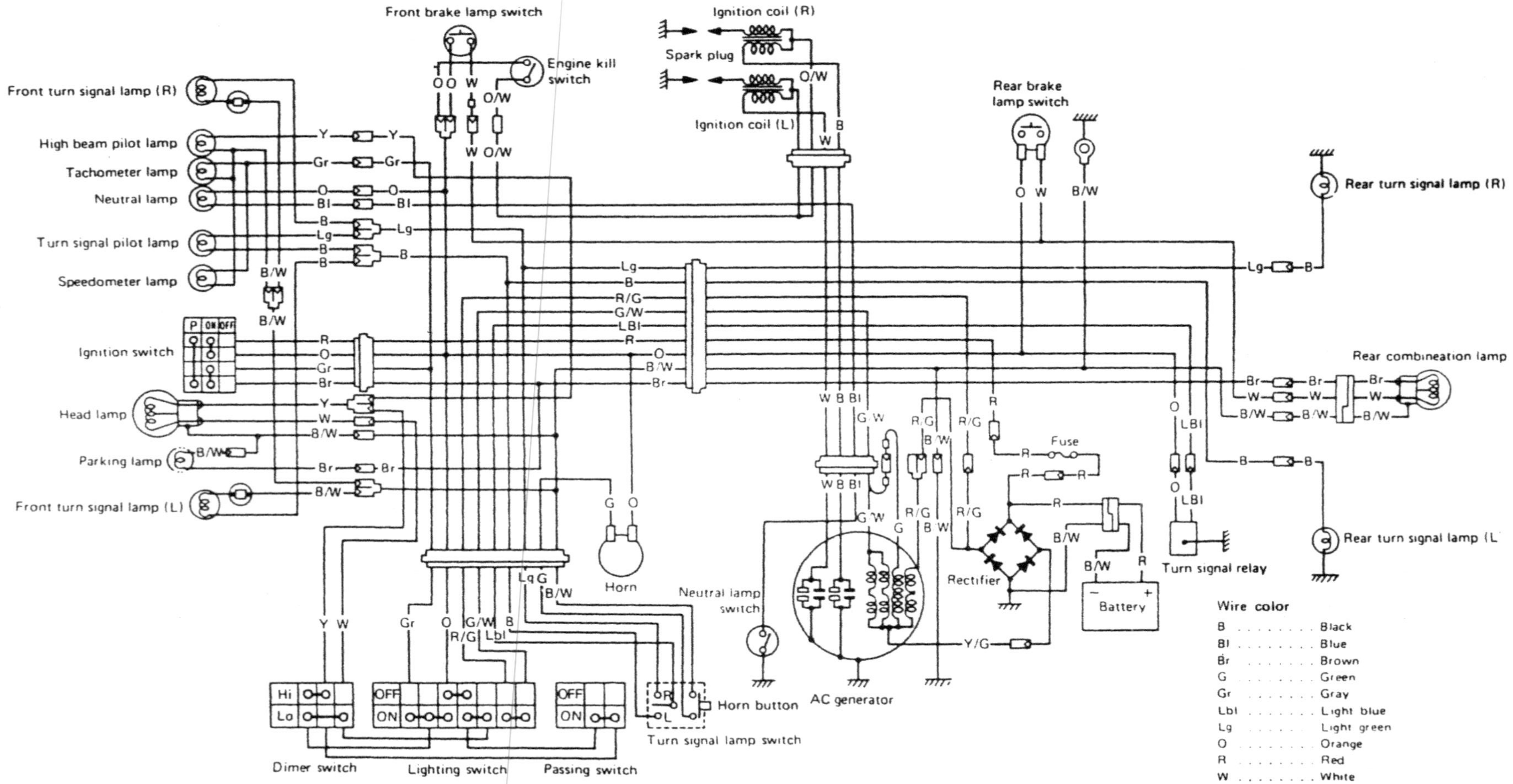

Rear turn signal lamp (R)

Rear combination lamp

Rear turn signal lamp (L

Wire color

B	....	Black
Bl	....	Blue
Br	....	Brown
G	....	Green
Gr	....	Gray
Lbl	....	Light blue
Lg	....	Light green
O	....	Orange
R	....	Red
W	....	White
Y	....	Yellow
B/W	....	Black with white tracer
G/W	....	Green with white tracer
R/G	....	Red with green tracer
Y/G	....	Yellow with green tracer

Turn signal relay

Battery

Fuse

Rectifier

AC generator

Horn button

Neutral lamp switch

Turn signal lamp switch

Rear brake lamp switch

Ignition coil (R)

Spark plug

Ignition coil (L)

Engine kill switch

Front brake lamp switch

Horn

Passing switch

Lighting switch

Dimer switch

Front turn signal lamp (R)

High beam pilot lamp

Tachometer lamp

Neutral lamp

Turn signal pilot lamp

Speedometer lamp

Ignition switch

Head lamp

Parking lamp

Front turn signal lamp (L)

GT 250A Wiring diagram — general

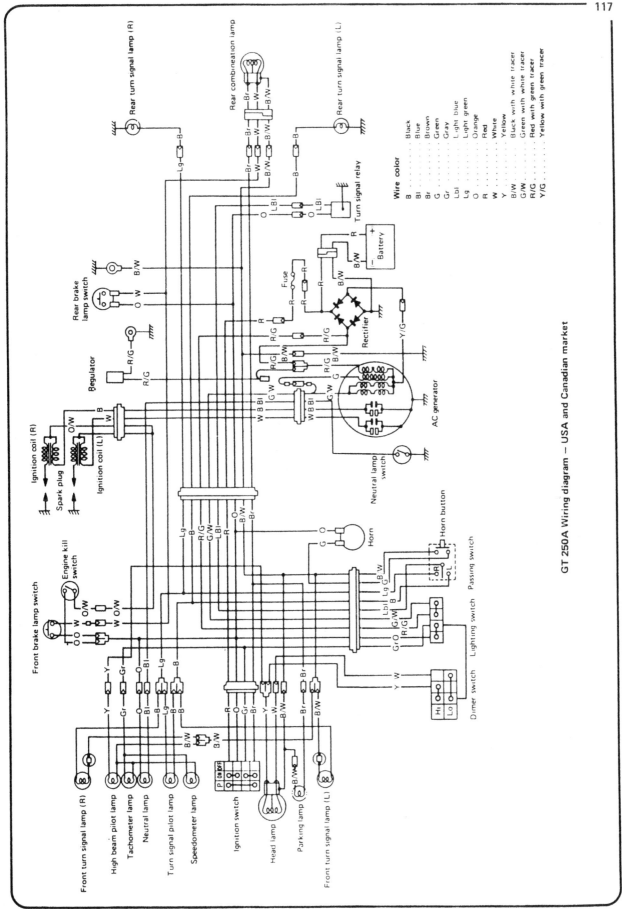

GT 250A Wiring diagram – USA and Canadian market

English/American terminology

Because this book has been written in England, British English component names, phrases and spellings have been used throughout. American English usage is quite often different and whereas normally no confusion should occur, a list of equivalent terminology is given below.

English	American	English	American
Air filter	Air cleaner	Mudguard	Fender
Alignment (headlamp)	Aim	Number plate	License plate
Allen screw/key	Socket screw/wrench	Output or layshaft	Countershaft
Anticlockwise	Counterclockwise	Panniers	Side cases
Bottom/top gear	Low/high gear	Paraffin	Kerosene
Bottom/top yoke	Bottom/top triple clamp	Petrol	Gasoline
Bush	Bushing	Petrol/fuel tank	Gas tank
Carburettor	Carburetor	Pinking	Pinging
Catch	Latch	Rear suspension unit	Rear shock absorber
Circlip	Snap ring	Rocker cover	Valve cover
Clutch drum	Clutch housing	Selector	Shifter
Dip switch	Dimmer switch	Self-locking pliers	Vise-grips
Disulphide	Disulfide	Side or parking lamp	Parking or auxiliary light
Dynamo	DC generator	Side or prop stand	Kick stand
Earth	Ground	Silencer	Muffler
End float	End play	Spanner	Wrench
Engineer's blue	Machinist's dye	Split pin	Cotter pin
Exhaust pipe	Header	Stanchion	Tube
Fault diagnosis	Trouble shooting	Sulphuric	Sulfuric
Float chamber	Float bowl	Sump	Oil pan
Footrest	Footpeg	Swinging arm	Swingarm
Fuel/petrol tap	Petcock	Tab washer	Lock washer
Gaiter	Boot	Top box	Trunk
Gearbox	Transmission	Two/four stroke	Two/four cycle
Gearchange	Shift	Tyre	Tire
Gudgeon pin	Wrist/piston pin	Valve collar	Valve retainer
Indicator	Turn signal	Valve collets	Valve cotters
Inlet	Intake	Vice	Vise
Input shaft or mainshaft	Mainshaft	Wheel spindle	Axle
Kickstart	Kickstarter	White spirit	Stoddard solvent
Lower leg	Slider	Windscreen	Windshield

Index